HEIDELBERG SCIENCE LIBRARY | Volume 15

Vegetation of the Earth

in Relation to Climate and the Eco-Physiological Conditions

Heinrich Walter

Translated from the second German edition by
Joy Wieser

Springer-Verlag New York Heidelberg Berlin

Heinrich Walter
o. Prof. em. Dr. h. c.
University of Hohenheim
Stuttgart-Hohenheim
Germany

Title of original German edition: Vegetationszonen und Klima.
Publisher: Verlag Eugen Ulmer, Stuttgart.

Library of Congress Catalog Card Number 72-85947.
Second printing, March 1975
Third printing, March 1977
Printed in the United States of America.

ISBN 0-387-90046-2 Springer-Verlag New York · Heidelberg · Berlin
ISBN 3-540-90046-2 Springer-Verlag Berlin · Heidelberg · New York

Preface

Ecology is current, exciting, relevant, and offers guides to action, even some hope of harmony, order, wise use, a congenial environment for mankind in an overpopulating world. Plant ecology is basic to general, animal, systems, paleo-, and human ecology. Plants are the primary producers. They dominate the flow and cycling of energy, water, and mineral nutrients within ecosystems. The structure of the vegetation which plants form determines much of the character of the landscapes in which other organisms live and prosper, including men and women.

Plants are immediately at hand for study. They are evident, mobile only in certain stages, familiar, easily identified and related to a rich literature on the properties of various kinds of plants. If we know why plants grow where they do, we know a good deal about why organisms other than plants live where *they* do.

Plant ecologists need a general botanical background. Professor Walter has that background. His multivolume textbook series, "Introduction to plant science", includes books on general botany, systematics, ecology in a strict sense (2nd ed., 1960), plant geography (2nd ed. with H. Straka, 1970), and on vegetation. Ellenberg covers the last topic with books on principles of vegetation organization (1956) and on the vegetation of central Europe (1963). These texts are in German, published by Ulmer (Stuttgart), and the reader can be referred to reviews of those on ecology in a broad sense in Ecology 38(4) : 666-8, 1957; 43(2) : 346, 1962; 47(1) : 167-8, 1966; and J. Ecology 55(1) : 234-5, 1967.

It is not enough for the plant ecologist to be well-grounded in principles. Our so-called ecological principles need continual re-examination, questioning, testing. Principles must be drawn from, and applied to, the specific ecological relationships of plants in particular ecosystems. Concrete ecosystems are the testing ground for principles as well as their source. So the ecologist needs an overview of the earth's plants and its vegetation. What are the possibilities and actualities of plant growth and vegetation organization?

Professor Walter's two volumes in German on "Vegetation of the earth considered eco-physiologically" (1964 and 1968) are an admirable summary of much of this basic plant ecology. They are separated by a language barrier from many persons who need to use them. They have been justly praised by A. Löve (Ecology 50(6) : 1105-1106, 1969) and by Grubb (J. Ecology 58(1) : 315-316, 1970) [1]. They record Walter's very extensive first-hand knowledge of much of the earth's vegetation. They add the progress made in our ecophysiological understanding of how and why plants grow where they do throughout the world since Schimper's founding work, "Plant geography on a physiological basis" (1898, 1903 in English, 2nd German edition by von Faber in 1935). Walter's books are indispensable to every practicing plant ecologist, as Löve says. Their 1593 pages are a rich feast. They also stand apart, separated, by their awesome scholarship, bulk, cost, their rich detail. Many students who lack a good geographical schooling or have only a rudimenary plant taxonomic background find them difficult to digest. Further, Walter's books are traditionally scientific, international in intent and scope, but this tradition contrasts with a wide spread general retreat into parochialism which sometimes seems to be a corollary of the potentially wide intellectual scope of ecology. Or perhaps we all just have too much to do.

In 1970 Professor Walter published a small volume on the ecology of zonal kinds of vegetation, in relation to climate, viewed causally, and covering all the continents. The English edition of that book you have in your hand. The German edition was reviewed in Ecology 52(5) : 949, 1971. The book is short but replete with facts. It places these facts in a consistent frame of reference. It suggests where more factual data are needed. It is also neat, precise, very readable, an excellent summary of two larger books, and current.

In recent years eco-physiology has attracted much interest from very skilled botanists. Technical progress in instrumentation has made possible some accurate measurements of photosynthesis and transpiration in the field, rapid chemical analyses of soils and plants, the sensing of elements of the heat balances of organisms, measurements of water potential in both soils and plants, computer modelling of the photosynthetic and respiratory processes in stands of vegetation,

[1] English edition of Vol. 1 cf. J. Ecology 60(3) : 940-941, 1972.

summarization of masses of data which would have formerly over-whelmed the investigator, etc. No synthesis of the principles of eco-physiology has appeared. But Walter's short book does something else that is very valuable. It provides a framework of ecological descriptions of kinds of zonal vegetation into which ecophysiological data must fit.

Plant eco-physiology can be used variously. Plant success, adaptations to environment whether ecotypic or plastic, should be definable in eco-physiological terms. For others the goal may be understanding the physiology of vegetation, those organizations of individual plants which many botanists ignore. Progress has been made in studying the physiology of crop stands. Walter records in this book new data on biomasses which summarize the results of the physiological processes in various kinds of vegetation and express their structure.

Under the International Biological Program many kinds of ecosystems are being studied. In the U.S. these include grasslands, deserts, deciduous, coniferous, and tropical forests, and tundra. A project on "Origin and structure of ecosystems" says, "The fundamental biological question that this program is asking is whether two very similar physical environments acting on phylogenetically dissimilar organisms in different parts of the world will produce structurally and functionally similar ecosystems. If the answer is no, there cannot be any predictive science of ecology. In effect, knowledge acquired from studying a given ecosystem cannot be applied to an analogous ecosystem, unless similar physical environment indeed means similar ecosystem" (U.S. National Committee for the International Biological Program. National Academy of Sciences, Washington, D. C. Report 4 : 46). The answer to the question is obviously no in general, but the conclusion suggested does not follow. True, there are many similarities between the plants and vegetations in different parts of the world that have similar climates, soil parent materials, topographies, fire histories, and plant successional and soil developmental histories even though their biotas differ, and Walter mentions examples repeatedly. But he also mentions exceptions. Ecology is not a simple matter of adaptations to environment. It must consider that biotic diversity over the earth is an ecological factor, and Walter outlines some of the classical conclusions on floristic diversity in the first pages of his text. The evidence is already clear that the partial effects of differing genetic inputs into environmentally closely similar ecosystems persist ecologically in the form of specific structures and functions. The C-4 photosynthetic pathway is a splendid, so far almost ecologically meaningless, example.

Further, structure and function should be considered separately. They have no necessary and invariant connection. Premature correlation, such as Schimper's bog xeromorphosis, is a mistake. Walter' generalization of "peinomorphosis" represents real progress. While the apparent congruences of structure and function are striking, study of them in detail and the evident exceptions have led to solution of many ecological puzzles.

Approaching the problem from another direction, ecological principles are applied widely in such kinds of land management as agriculture, forestry, range and pasture management, pollution control, park and wilderness management, natural area maintenance. Ecological principles evidently exist and do allow predictions. Walter's little book is a good corrective to the kind of hurried, mistaken generalization that was quoted.

Of course evolution is a dimension of ecology. Unfortunately the concept of adaptation is only used by evolutionists and not often measured. In fact, there is often a shifting from form to function and vice versa in discussions of adaptation, a substitution of two unknowns for one. Walter does not make this mistake either. He does make clear how much more research needs to be done to describe plant and vegetation structure and function, their correlations, and their relationships to environment. And his book records a fine start.

Professor Walter is a plant ecologist of very wide and long expierience. He and his wife are knowledgeable, enthusiastic, indefatigible, helpful, field companions. His investigations, and those of his students, have suggested, checked, and documented his ecological ideas derived from extensive travel and residence in all the continents. His teaching has clarified their presentation.

Many have questioned whether plant ecology in its century of development since Haeckel coined the parent term in 1866 has developed any principles. This book provides an affirmative answer.

University of California, Davis Jack Major
February 8, 1972

Foreword

Because botany is becoming more and more a laboratory science, with molecular biology and biochemistry in the forefront of interest, it is more important that students not neglect the worldwide and equally fascinating problems of the ecophysiological aspects of field botany on a continental scale. The population explosion and subsequent urbanization, as well as the ever-increasing disruption of nature including pollution of air and water, currently present the greatest dangers to the physical and psychical well-being of mankind. For this reason it seemed to be appropriate to make available an abridged form of the author's two-volume *Vegetation of the Earth*, Jena-Stuttgart (1964, 1968), which deals comprehensively with the problems touched upon in this survey. Voluminous data in the form of tables and diagrams, as well as sources of literature, are to be found in the larger book (third edition of Volume I, in press). Only a few recent or as yet unquoted references have been included at the end of this book in the form of notes. The as yet unpublished results of the author's most recent journeys (Venezuela, 1968 and North America, 1969), made possible by the kind support of the Universidad Central of Caracas and the Ecology Center in Logan, Utah, have been dealt with briefly in summarized form (Fig. 1).

Ecophysiology, which aims at ascertaining the factors responsible in nature for preserving the innermost integrity of the plant world, forms the basis of studies concerning the maintenance of a healthy environment for modern man. It is thus fundamental for environmental studies, for nature conservancy and for the problems of the underdeveloped countries, insofar as they involve utilization of the plant cover. Physiological experiments provide a useful supplement to ecophysiology, but they cannot replace thorough ecophysiological investigations carried out under natural, field conditions. Successful ecophysiological studies require familiarity with all aspects of competition between plants.

The sources of the figures are to be found on p. 231.

Stuttgart-Hohenheim   Heinrich Walter
December 25, 1969

Contents

Vegetation of the Earth

Introduction

1 Floristic Realms

The vegetation of the Earth as we know it today is the result of a long process of development under the influence of environmental factors, both past and present. The positions of the continents relative to each other and to the poles have repeatedly changed during the course of time so that the floristic development in the various parts of the Earth has followed divergent routes. These historical events have led to a differentiation into floristic realms (1).

For example, of the phylogenetically relatively old group of conifers, the family Podocarpaceae and the genus *Araucaria* occur only in the Southern Hemisphere, while the large family Pinaceae and nearly all Taxodiaceae occur in the Northern Hemisphere; the Cupressaceae, however, are to be found scattered over all continents.

The distribution of flowering plants, or angiosperms, the youngest branch of the plant kingdom, is much more sharply differentiated. The oldest families of this plant group are known to have existed in the lower Cretaceous period, but their main development took place in the Tertiary period, when the land masses had already split up into the different continents. In the Northern Hemisphere, however, this held true to a limited extent only, since it was not until the Pleistocene epoch that North America and Greenland finally separated from Eurasia. As a result, floristic differences in this area are so small that these continents can be considered as one floristic realm, the *Holarctic*. Much larger differences exist, however, between the tropical floras of the Old and New Worlds, so that two floristic realms must be distinguished, the *Palaeotropic* and the *Neotropic*, respectively. Floristically, the southernmost parts of South America and Africa, and Australia in its extreme isolation, have still less in common. For this reason, three floristic realms have been distinguished: the *Antarctic*, which comprises the southern tip of South America and the subantarctic islands; the *Australian*, which is geographically identical with

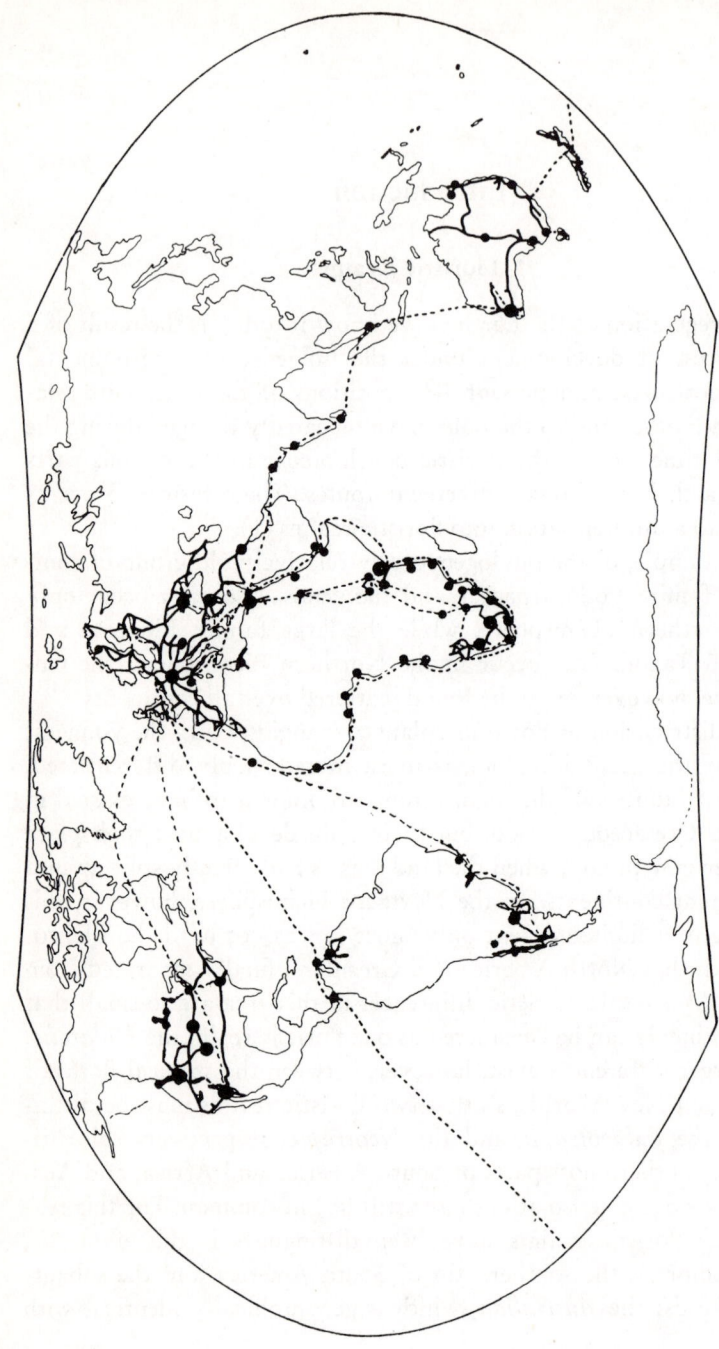

Fig. 1. The author's research journeys, by car or train, by sea or air; large dots indicate longer visits to research institutes. The Russian literature has been consulted for information concerning the inaccesible regions of Asia.

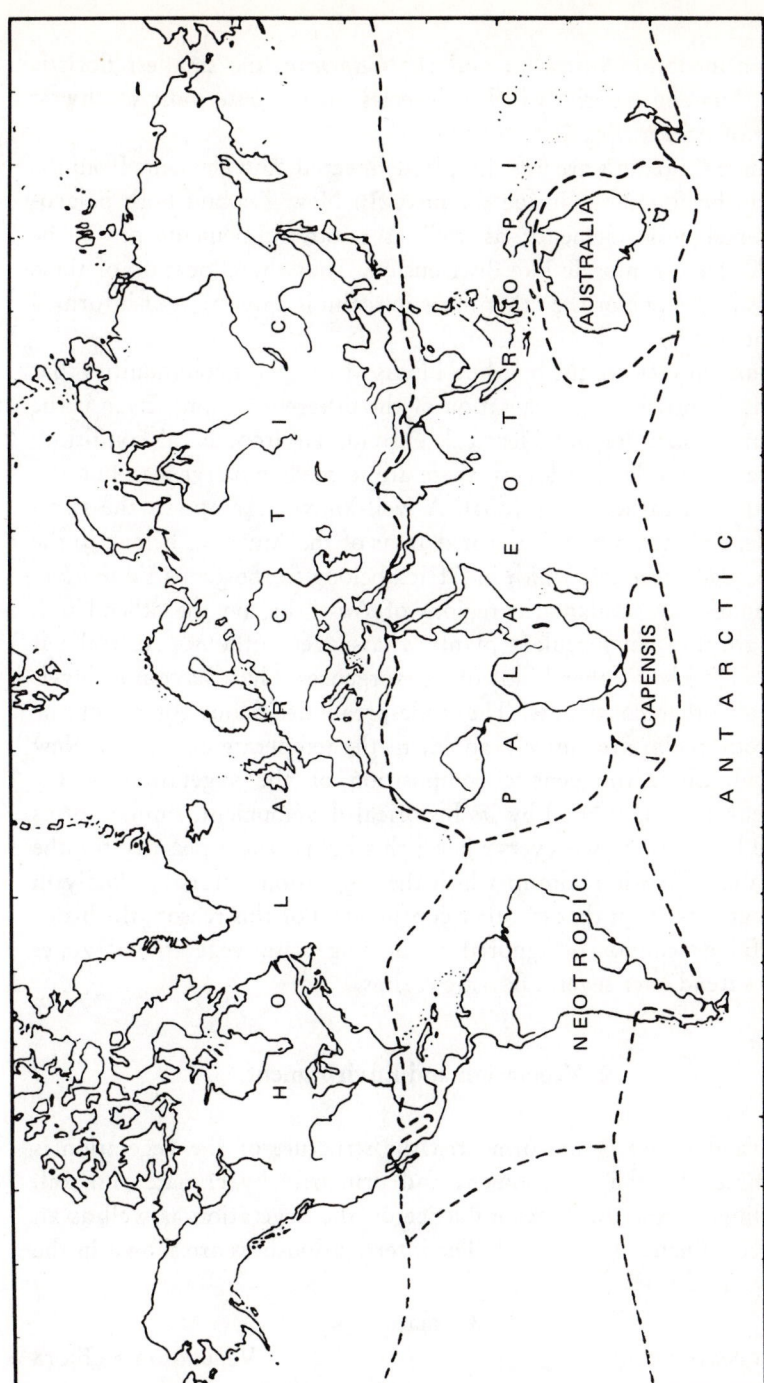

Fig. 2. The floristic realms of the Earth (Diels and Good, modified, from Walter/Straka 1970). In New Zealand and Tasmania antarctic, as well as palaeotropic and australian elements occur.

the continent of Australia; and the *Capensic*, the smallest floristic realm, but one especially rich in species, in the outermost southwest corner of Africa (Fig. 2).

These six realms are not sharply delineated, and elements from the one can be found far inside the next. In New Zealand both palaeo-tropic-melanesic elements, as well as antarctic elements are to be found, often in mosaic-like distribution. Thus the allocation of these islands to the one or the other floristic realm is a question of informed judgment.

Plant species are the building blocks of the plant communities that together constitute the vegetation of the different regions. Even if the building blocks are not identical, extreme environmental conditions can lead to similar life forms; these are termed *convergences*, but are, nevertheless, rather exceptional. A well-known example is the stem-succulent plants, which in the arid areas of the Americas belong to the family Cactaceae but which in Africa belong to the genus *Euphorbia*. In climatically similar arid regions of Australia, on the other hand, there are no stem succulent plants whatsoever, although Australia is especially rich in other kinds of convergences which have not developed on other continents. The widespread deciduous forests of the Holarctic realm are entirely absent in the temperate climate of New Zealand. Since the genetic composition of the vegetation of the different areas is limited by its historical development, similar forms of life have not arisen everywhere; this holds true especially for the Australian floristic realm in which the vegetation differs markedly in physiognomy from that of other continents. For this reason, the historical factor cannot be ignored in dealing with vegetational zones which extend over several floristic realms.

2 Vegetation and Environment

In a floristically uniform area the structure of the vegetation is determined by the environment and primarily by climate and soil. The climate exerts a direct influence on the vegetation as well as an indirect influence via the soil. The interrelationships are shown in the following scheme:

Parent rock → Soil   Climate   Vegetation ← Flora

The nature of both soil and vegetation is determined by the climate, although the type of parent rock influences the soil and the flora is of significance for the vegetation. Soil and vegetation are so closely interrelated that they may be considered together almost as an entity. Moreover, since both affect the climatic conditions in the air layer nearest the soil, they can also be said to influence the microclimate. The environment of plants is definable as the sum of the factors acting upon them, the physicochemical factors (without competition) being termed the habitat and the locality of growth being known as the biotope. The factors influencing the habitat are often listed as climatic, orographic, and edaphic (soil), but this is of little help from the ecological point of view. The important factors should rather be listed as follows:

1. Heat (temperature)
2. Water (hydrature)
3. Various chemical factors
4. Mechanical factors

It is immaterial to the plants whether favorable thermic conditions are provided by the regional climate or are due, for example, to a sheltered biotope on a southern slope. Similarly, the required soil moisture may result from favorable atmospheric precipitation or may be secured by limited evaporation on a northern slope or may be due to the soil structure and proximity of ground water. What is important is that the plants suffer no lack of water.

For the differentiation of the plant cover into vegetational zones, the temperature and moisture conditions at ground level are decisive. They are in general determined by the climate (p. 17). In all biological processes external factors are only of direct importance insofar as they influence the state of the living protoplasm.

In considering thermic conditions, it is the temperature at which the vital processes take place in the protoplasm which is significant. Animal organisms may be termed cold-blooded, or poikilothermic, if their body temperature, and hence the temperature of the protoplasm, is dependent upon the external temperature and varies with it; they may be termed warm-blooded, or homoiothermic, if they are to a large extent independent of the environmental temperature and have a more or less constant body temperature. For the warm-blooded organism it is therefore meaningless to measure the external temperature and relate it to the course of physiological processes.

Plants are invariably *poikilothermic*. The temperature of the surrounding air thus provides a clue to the temperature of the protoplasm. Certain minor deviations of a purely physical origin are induced by strong radiation and must be determined in an accurate ecophysiological investigation. However, it is usually considered sufficient to record the air temperature.

With respect to their water economy, plants behave in as complicated a manner as do animals with respect to temperature; a distinction must therefore be made between *poikilohydric* and *homoiohydric* plants (2).

Protoplasm is physiologically active only when its water content is high, that is to say, when it is in a hydrated or swollen state. If the cells dry out, the protoplasm falls into a latent condition and exhibits no detectable signs of life, or it dies. According to the thermodynamics of imbibants the degree of hydration depends upon the relative activity of water (a), where $a = p/p_0$, that is to say it is equivalent to the relative water-vapor tension. Expressed in percent (pure water = 100 percent), this is termed "hydrature," which, in turn, is the same as the air humidity (percent r.). Since the vital processes are to a large extent dependent upon the degree of hydration of the protoplasm, it is essential to know the hydrature of the protoplasm.

In such *poikilohydric organisms* as the lower terrestrial plants (bacteria, algae, fungi, and lichens), the hydrature depends entirely upon the humidity of the surrounding air. If the plants are in contact with water or if the surrounding air is saturated with water vapor, then the protoplasm of such species is hydrated and active to an almost maximal degree. In dry air, however, dehydration takes place, and the protoplasm passes over into a resting condition without dying. The cells of such organisms have no vacuoles, or very small ones, so that the changes in volume during desiccation are small and the protoplasmic structure remains undamaged. The lower limit of hydrature (atmospheric humidity) at which growth is still demonstrable is very high for most bacteria, is high but varied for unicellular algae and moulds, and in only a few does it sink to *70 percent, a value which corresponds to the extreme lower level of hydrature compatible with any sign of life* (2). The productivity of these poikilohydric organisms is small, and they contribute but little to the entire terrestrial phytomass. For this reason they have, until now, been afforded little attention, although they are often of much more widespread distribution, particularly in deserts, than is generally supposed.

The terrestrial *homoiohydric plants* are of much greater importance and include all the Cormophytes which developed originally from green algae. Their cells have a large central vacuole, and since the hydrature of the surrounding protoplasm in the cell is in equilibrium with that of the vacuolar cell sap, it is thus not so immediately dependent upon water conditions outside the cell. The entire vacuolar cell sap, termed here the *vacuome,* constitutes an internal watery milieu for the higher plants. It is this "internal milieu" which, in the course of phylogenetic development, has made possible a transition from an aquatic to a terrestrial way of life as well as a steadily improving adaptation to arid conditions. As long as terrestrial plants are able to keep the cell-sap concentration of the vacuome low, the protoplasm remains highly hydrated. In other words, the hydrature of the protoplasm remains high, independent of the moisture in the surrounding air. Since, however, the prerequisite for photosynthesis of autotrophic plants is gaseous exchange with the atmosphere, which, in turn, leads to water losses due to transpiration, a complicated system has been evolved to serve these ends; roots for water uptake, vessels for water transport, and stomata for the regulation of transpiration. This apparatus is but imperfectly developed in mosses, and they are therefore generally confined to very damp habitats. The ferns, too, with a somewhat inefficient transport system consisting solely of tracheids, avoid dry habitats. Those mosses and the few ferns and Selaginella spp that have penetrated into desert regions have had to change to a *secondarily poikilohydric way of life* in order to survive dehydration during times of drought. They achieved this ability to withstand desiccation, not normally possessed by plants with strongly vacuolated cells, either by a diminution of cell size with reduction of the vacuoles or else by vacuolar storage of colloids (tannins). These tannins solidify upon even a minute loss of water and so prevent deformation and damage to the protoplasm. With respect to water economy, the most effective adaptation to a terrestrial way of life has been achieved by the Angiosperms which have even penetrated into extreme desert regions. Measurement of their cell-sap concentration shows that, without too greatly checking the gaseous exchange necessary for photosynthesis, they are able to keep the concentration of the cell sap low and thus maintain the hydrature of the

7

cytoplasm at a high level. A rise in the concentration of the cell sap (decrease in osmotic potential) and a resultant dehydration of the cytoplasm is not an appropriate adaptation for desert plants, although this statement is still to be found in text books. It is rather an indication of a disrupted water balance and represents a threat to survival.

The measurement of such external factors as precipitation, humidity, or water content of the soil offers little information concerning the water activity in the protoplasm (hydrature and degree of hydration), just as the measurement of external temperatures offers little information in studying warm-blooded homoiothermic animals.

Determination of the cell-sap concentration in atm (or osmotic potential), which is directly related to the relative water vapor tension (hydrature), *is the only means of ascertaining whether the plant has been affected (dehydration of the protoplasm) by alterations in external conditions, especially by drought.* Measurement of the suction tension (water potential) is much more difficult to carry out than the measurement of cell-sap concentration and is, indeed, much less relevant. In describing the habitats of the various vegetational zones we shall, therefore, in addition to giving the usual data concerning external factors, indicate, where possible, the cell-sap concentration and its variations. This gives an idea as to the hydrature of the protoplasm and is of particular interest in discussing arid districts, where the water factor is of paramount importance.

These explanations are necessary because textbooks, in dealing with water economy, make no distinction between poikilohydric and homoiohydric plants.

The importance of light and of chemical and mechanical factors may be assumed familiar in general outline, and special points of interest will be dealt with in the individual sections.

3 Competition in Plant Communities

The widespread assumption that the distribution of plant species is directly dependent upon the physical conditions prevailing in the habitat is incorrect; they are of *only indirect importance, insofar as they influence the competitive power of the various species.* Only at the absolute distribution limit, in arid or icy deserts, on the edge of the salt desert, or where the forest shade is at its deepest, are the physical environmental factors (usually one particular, extreme factor) of direct importance. Apart from such exceptions, plant species are capable of existing far beyond their natural distributional areas if they are protected from competition with other species. The northeastern limit of distribution of the European beech *(Fagus sylvatica)*, for example, runs through the Vistula region of Poland, although beech is found growing far to the north and southeast of this limit in the botanical gardens in Helsinki and Kiev. *The natural limit of distribution of a particular species is reached when, as a result of changing physical environmental factors, its ability to compete, or its competitive power, is so much reduced that it can be ousted by other species.* It depends, therefore, also upon the presence of certain competitors. For the beech, these are the hornbeam *(Carpinus betulus)* on the eastern boundary, the oak *(Quercus robur)* to the north, and the spruce *(Picea abies)* in the mountainous regions. That the northeastern beech limit takes a similar course to the January isotherm for $-2°$ C, that the northern limits of oak distribution follow the line indicating four months of the year above $+10°$ C, and that the northernmost spruce boundary coincides with the July isotherm of $+10°$ C do not necessarily indicate a direct causal relationship. It can, at most, be concluded that in the case of beech the increasingly cold winter toward the east, and for oak and spruce the shorter summer to the north, radically reduce the competitive power of these species.

The conditions under which a species occurs most abundantly in nature may be termed the *ecological optimum,* and the conditions under which it thrives best in the laboratory (phytotron or growth chamber) may be termed the physiological optimum. These optima, however, are rarely identical, as may be seen from Fig. 3.

The distribution of a species, therefore, is not an absolute guide to its physiological requirements. The fact that, for example, in western Europe the Scotch pine *(Pinus sylvestris)* is found under natural conditions only on dry or calcareous slopes or on acid, boggy soil, is due to its having been supplanted from a more suitable habitat by stronger competitors. The western North American lodgepole pine *(Pinus contorta)* or the eastern jack pine *(Pinus banksiana)* behave similarly. Therefore, the knowledge acquired in the phytotron about the physiological needs of a species forms an insufficient basis either for a prediction of its distribution in nature or for an understanding

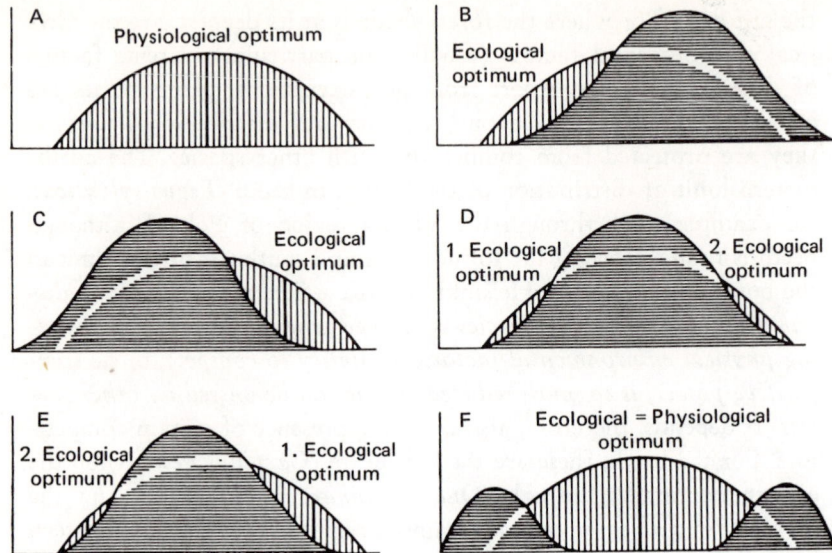

Fig. 3. Growth curves (vertical shading) of one species without (A) or under pressure due to competition (B — F, horizontal shading). Ordinate: Growth intensity and production of organic matter; abscissa: variable habitat factors.

thereof. Whether or not a species colonizes a habitat which is physiologically suitable depends, apart from the historical factor, mainly upon the nature of its competitors.

Competition is generally said to take place when the growth or development of a species is unfavorably influenced by the presence of another, without the occurrence of parasitism. It is nearly always in action wherever several species occur close together and no form of dependence exists, and it can be recognized when a particular plant develops more luxuriantly in isolation than when in a plant community. Inhibition in the course of competition results mainly from the cutting off of light by surface organs or from lack of water or nutrients due to root competition. Whether, apart from these factors, an important role in the competitive struggle can be attributed to certain specific substances excreted by the plants themselves (allelopathy) under natural conditions, has not yet been satisfactorily proved.

A distinction must be made between *intraspecific* competition, occuring between individuals of the same species, and *interspecific* competition, taking place between individuals of different species.

The former type of competition eliminates the weaker individuals and helps to preserve the species. In interspecific competition a species can achieve dominance and supplant others or, in a mixed population, a state of equilibrium may be established based on the competitive power of the individual partners. In mountainous regions, for example, it can be seen that, at the beech-spruce boundary, the beech is absolutely dominant on southern slopes and spruce on the northern slopes, while on eastern and western slopes the two are fairly well balanced and a mixed population is formed. This can also happen if, as seems to be the case in tropical rainforests, the seedlings of a certain species develop better beneath other species than beneath individuals of the same species.

The competitive power of a species is a phenomenon of great complexity and *is only definable for a given set of environmental conditions.* The entire morphological and physiological properties of a species are of significance in this respect. Biennial species are competitively more powerful than annuals owing to the fact that they commence their second year of growth with the large reserves accumulated during the first. For the same reason, perennial herbs are superior to biennials from the second year onwards. Woody species win out against perennial herbs if they have not been suppressed during their early years and have succeeded in producing ligneous axial organs to raise them above the herbage layer.

As a result of competition, similar combinations of plant species occur repeatedly on similar habitats within a limited area and are termed *plant communities.* Examples in central Europe are beech woods together with their herbaceous flora, flood-plain forests, meadows or pastures, various types of bogs or fens, and so on.

In a stable plant community the different species are in a certain state of ecological equilibrium with each other and with their environment. The following factors are of importance for this equilibrium:

1. interspecific competition

2. the dependence of one species upon the presence of others (e. g., shade species).

3. the occurrence of species that complement one another either spatially or temporally so that every ecological niche is filled.

The natural community is thus saturated and invading species can gain no foothold, although they are much more successful if the state of equilibrium is disturbed.

The equilibrium of a plant community is dynamic rather than static in that some individuals die off, while others germinate and grow. At the same time the individual species are continually exchanging places. Quantitatively, too, the species composition exhibits certain deviations since external conditions vary from year to year, rainy years follow upon dry ones, and so on. Consequently, sometimes one species will be favored, sometimes another. If the conditions in the habitat alter continously in one direction e. g., if the groundwater table slowly rises over many years then the combination of species also changes; some species disappear and others infiltrate from outside until finally a new plant community arises. Such a series of events is termed a *succession*. If the alterations in the habitat arise from natural causes and originate on parent rock then it is called a primary succession; it is usually a very slow process. More often a rapidly progressing so-called secondary succession results from human interference (draining of water meadows, deforestation, abandonment of fallow land, ceasing to mow meadows, etc.) or is caused by such catastrophes as hurricane winds, fire, etc. If human interference continues over a long period of time and is of a uniform nature, then an anthropogenic equilibrium develops. The plant communities arising are termed cultural formations if they are intensively utilized and semi cultural formations if more extensively utilized. They constitute the vegetation in areas densely populated by man.

4 Ecosystems or Biogeocenoses

A continuous energy turnover and cycling of nutrients take place in a plant community. The plants, together with the animal organisms and the inorganic environment, form *an ecosystem which, however, is not a closed system* since there is an inflow of external energy from solar radiation and of matter in the form of precipitation or from gaseous exchange, dust deposits, and so on. At the same time, energy is lost in the form of heat, and loss of matter occurs as a result of gaseous exchange or in drainage water. If the ecosystem comprises a definite, limited, and homogeneous community, such as a forest stand or a moor, then it can conveniently be termed a *biogeocenose*. This is an entity consisting of the plants and animals, the soil permeated by the roots, and the air layer into which the plants extend (Fig. 4). The

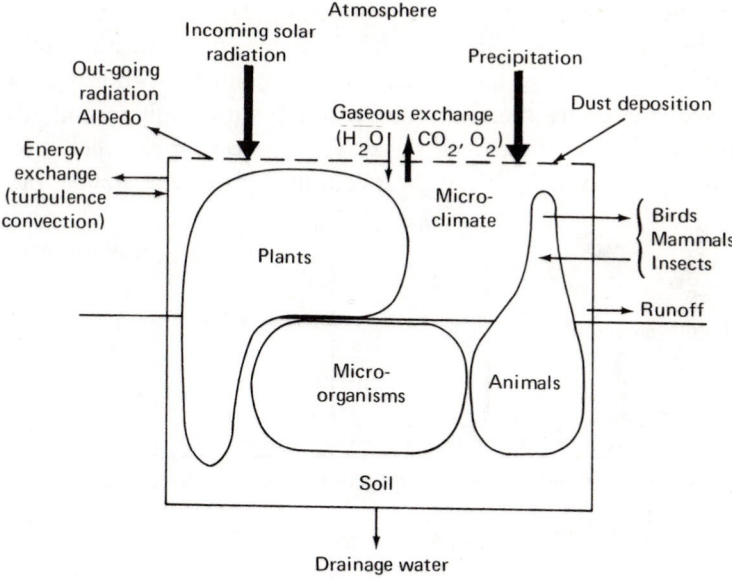

Fig. 4. Schematic representation of an ecosystem or biogeocenose (framed) in a state of exchange with the environment.

total dry plant matter of a biogeocenose is its *phytomass* and that of its dry animal matter its *zoomass*. According to their role in the biogeocenose, three groups of organisms may be distinguished:

1. *Producers*—the green plants, which in the process called photosynthesis are able to store the energy of sunlight as chemical energy by building up organic compounds from CO_2 and H_2O. This is the *primary production*.

2. *Consumers*—these are animal organisms which use plants as food and convert a small part of this matter into animal substance. This is termed *secondary production*.

3. *Decomposers*—to a large extent present in soil (bacteria, fungi, and protista). They break down or mineralize plant and animal remains into CO_2 and H_2O, thus completing the cycling of nutrients.

13

The cycling of carbon is linked up with certain *nutritional chains* beginning in the plant substance, as shown schematically in Fig. 5. Parallel with the cycling of nutrients is the *energy turnover*. Plants alone respire about half of the organic substance produced, whereby chemical energy is lost in heat production. This holds true for each step in the nutritional chain in which animal organisms and microorganisms participate until, when complete mineralization has been acieved, the entire chemical energy stored during photosynthesis is exhausted i. e., converted into heat. A biogeocenose can therefore be maintained only if light energy is continually made available. How-

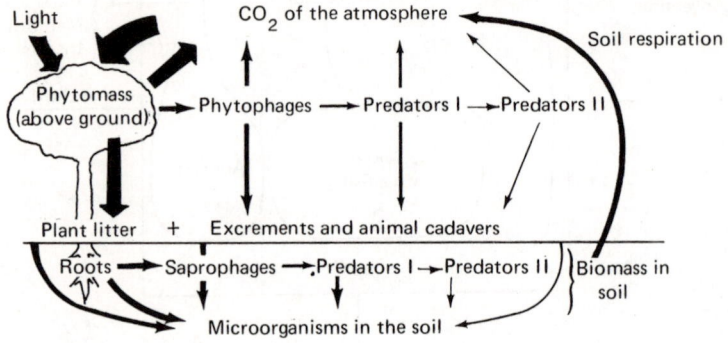

Fig. 5. Diagram of the food chain in a forest and the cycling of carbon, to which the energy turnover runs parallel.

ever, only a small proportion of the absorbed radiation is used for photosynthesis; the largest portion is needed for the evaporation of water during transpiration or is lost through heat emission or exchange. Ecophysiological research, which is still relatively young, aims at a quantitative and qualitative study of the cycling of nutrients and the energy turnover in the various biogeocenoses constituting the Earth's biosphere. An exact knowledge of primary production within the various climatic zones would be a big step forward, this forming, as it does, the basis of human nutrition.

A basic difference exists, nevertheless, between biogeocenoses such as virgin forest or natural steppe, that are in equilibrium with the environment, and those whose resources are utilized by man. In the former, the total mass of the green plants is completely mineralized by consumers and decomposers, and the nutrients set free are taken up again by the roots. In this way the phytomass (averaged over several years) remains constant, unless some catastrophe such as a storm or fire occurs. On the other hand, in a permanent forest managed by selective cutting, or a mowed grass prairie, and more especially in plantations or cultirated fields, part of the primary production is removed by man. Man could therefore be included among the list of consumers if it were not for the important difference that the mineral nutrients contained in the organic substance removed by him are lost to the biogeocenose and must be replaced by weathering of minerals or by fertilization if the equilibrium is not to be destroyed and a slow degradation of the community is not to be the consequence.

The terms "gross production" and "net production" appear often. "Gross production" refers to the entire organic mass arising from the photosynthetic activity of the green leaves of a plant community, including the portion that is immediately respired, whereas the term "net production" is used in a variety of senses. It is sometimes used to indicate the gross production less the mass respired by leaves, axial organs, and roots, and it is then identical with primary production. In forestry, however, net production is taken to mean the annual increase in wood, often, indeed, merely the increase in utilizable timber. "Net production" can also be used to denote the annual increase in phytomass, which in balanced, stable communities (on a large scale) is equal to zero. In speaking of net production, therefore, it is obviously essential to define exactly the way in which the term is being used.

This book is intended as an ecophysiological study rather than as a description of the vegetation. Plant communities will be dealt with, in the main, as components of ecosystems and, as far as this is possible, their relations to the environment and to primary production will be discussed.

Plant communities are easier to investigate than the freely moving animal communities, which accounts for the more advanced state of our knowledge about them. Plant communities constitute, too, the most important part of the biogeocenose; zoomass is always much smaller than phytomass, and its contribution to total metabolism is seldom of importance.

As an example, a 120-year-old west European mixed deciduous forest has been chosen. Approximate values for dry biomass are given (from Duvigneaud 1962):

Aboveground biomass (1 t/ha = 100 g/m²)

Producers		Consumers	
Leaves	4 t/ha	Birds	1.3 kg/ha
Branches	30 t/ha	Large mammals	2.2 kg/ha
Trunk wood	240 t/ha	Small mammals	5.0 kg/ha
Undergrowth	1 t/ha	Predators	undetermined
Total	275 t/ha	Total	8.5 kg/ha
		plus invertebrates	quantity unknown

Underground biomass	
Root mass	many tons per hectare
Soil fauna	1 t/ha, of which 0.2 t/ha earthworms
Soil flora	0.3 t/ha, made up of (number per g soil): Bacteria (95×10^6), actinomycetes (36×10^6), and fungi (1×10^6).

The quantity of humus ranges from 1 to 70 t/ha.

Half of the gross production is immediately lost by plant respiration. The remaining net production (primary production) amounts to 12 t/ha dry mass per annum, made up as follows: leaves of the trees 4 t/ha; production of aboveground wood mass 5 t/ha, of roots about 2 t/ha; and of undergrowth 1 t/ha.

5 Climatic Zones of the Earth

a Survey

It has already been emphasized that the heat or temperature factor and the water or hydrature factor, and, above all, their seasonal variations, are particularly important climatic elements. In general, temperature decreases from the equator toward the poles, while the seasonal variations become increasingly large. In tropical regions precipitation takes the form of the so-called equinoctial rains, which occur at the season when the sun is at zenith at midday. In the temperate zones the rainfall is cyclonic and is associated with low-pressure areas moving from west to east. These areas arise in the Northern Hemisphere in Iceland and the Aleutians, and are known as the Iceland Low and the Aleutian Low.

The resulting main climatic zones are:

I. *The equatorial zone*, lying roughly between 10° N and S, with a diurnal type of climate. This means that the diurnal variations in temperature are greater than the annual variations in the mean daily temperature, which is approximately 25—27° C throughout the entire year. Annual rainfall is usually high, the maxima occurring at the time of the equinox. Nevertheless, isolated arid regions may occur in this zone.

II. *The tropical zone*, lying to the north and south of I (to about 30° N and S). A certain seasonal variation in the mean daily temperature is already noticeable. Rainfall reaches a maximum at the time when the sun is at zenith, so that there is a rainy season in the summer and a dry season in the cool months. The duration of the latter increases as the distance from the equator becomes greater, and at the same time the annual rainfall decreases.

III. *The subtropical dry zone*, poleward of 30° N and S, in the region of the descending air masses, which get warmer as they descend and become very dry. Rainfall is very low, and the daytime temperatures are very high because of intense solar radiation. In the winter months, however, the temperature may sink to zero at night as a result of the greater net loss of heat energy in outgoing radiation. This is the hot desert zone.

IV. *The transitional zone with winter rain,* at latitudes around 40°. In summer it belongs to the high-pressure and dry-air zone, but in winter it receives cyclonic rain. It has a typical Mediterranean climate with no cold season, but with occasional frost and a long summer drought.

V—VIII. *The temperature zones* with cyclonic rain at all seasons, decreasing, however, as the distance from the ocean increases. A distinction must therefore be made between a wet, oceanic climate and a dry, continental climate with hotter summers and colder winters. The temperate climatic regions can be listed as follows:

V. *Warm temperate* climate, with scarcely any or no winter. It is extremely wet, especially in summer.

VI. *Typical temperate* climate, such as the central European or coastal northeastern U.S.A. with a cold, but not too long a winter, or with a winter almost free of frost and with very cool summers (extreme oceanic).

VII. *Arid temperate* climate, continental in character, with large temperature contrasts between summer and winter, and little precipitation.

VIII. *Boreal or cold temperate* climate with cool, wet summers and cold winters lasting more than six months.

IX. *The arctic climatic zone,* with low precipitation distributed over the entire year. Owing to the low temperatures there is only a short, wet nightless summer and a very long, cold, dark winter.

There are, however, certain marked deviations from this general zonation as, for example, the asymmetric conditions prevailing on the different sides of the equator. The climate of the Southern Hemisphere is generally cooler and more equable owing to the predominance of the ocean masses, and Zone III is little represented. The temperate zone occupies a small area; since only South America extends beyond 40° S, the boreal zone is entirely absent; and the Antarctic zone is almost solely represented by the icy Antarctic continent. South of 40° S a strong west wind blows continuously.

Further deviations in zonation can be attributed to the planetary winds: the trade winds blowing toward the equator bring rain only if they are intercepted by highlands, which also holds true for the monsoons blowing to the mainland of East Africa. On the other hand, rain is brought by the monsoons arising in the warm seas, such as the Guinea monsoon and the monsoon winds of India, Indonesia, and Southeast Asia.

Mountain ranges also lead to irregularities in zonation: Precipitation is high on the windward side on account of the "ascending air masses," but the lee side and the sheltered lowlands are dry.

The world distribution of precipitation, as shown in Fig. 6, is the product of all these factors combined with other, local influences.

b Climatic Diagrams

It is not the isolated climatic elements that are important to plants, but their combined and simultaneous effect. Climate must therefore be considered as a whole, with its characteristic seasonal weather pattern. This can better be achieved by means of *graphical representation* than by the use of indices and formulae. Climatic diagrams from 8 000 stations scattered over every continent are collected together in the "World Atlas of Climatic Diagrams" (Jena 1967).

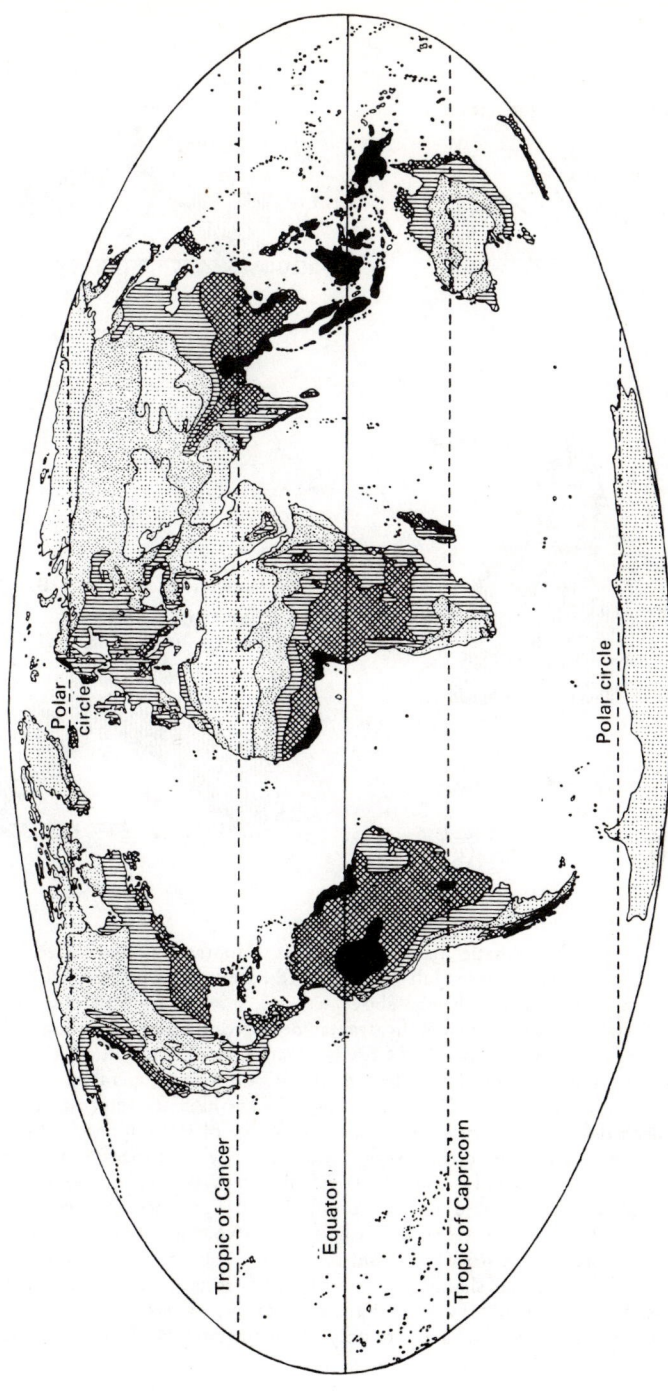

Fig. 6. Annual rainfall map. Thinly dotted = below 250 mm, thickly dotted = 250—500 mm, vertical shading = 500—1000 mm, cross-shading = 1000—2000 mm, black = more than 2000 mm.

19

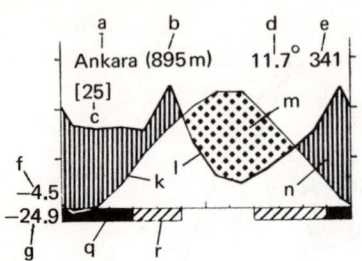

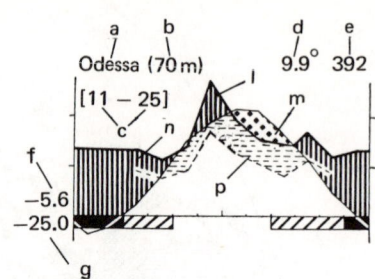

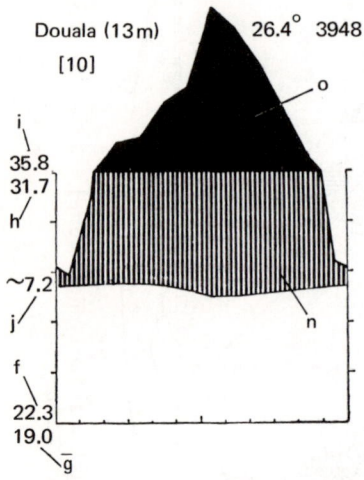

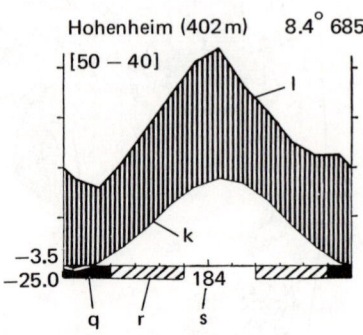

Fig. 7. Key to the climatic diagrams. Abscissa: Months (N. Hemisphere January—December, S. Hemisphere Juli—June); Ordinate: one division = 10° C or 20 mm rain. a = station, b = height above sealevel, c = duration of observations in years (of two figures the first indicatea temperature, the second precipitation), d = mean annual temperature in ° C, e = mean annual precipitation in mm, f = mean daily minimum of the coldest month, g = lowest temperature recorded, h = mean daily maximum of the warmest month, i = highest temperature recorded, j = mean daily temperature variations, k = curve of mean monthly temperature, l = curve of mean monthly precipitation, m = relative period of drought (dotted), n = relative humid season (vertical shading), o = mean monthly rain > 100 mm (black scale reduced to $^1/_{10}$), p = reduced supplementary precipitation curve (10° C = 30 mm) and above it (dashes) dry period, q = months with mean daily minimum below 0° C (black) = cold season, r = months with absolute minimum below 0° C (diagonal shading) = late or early frosts occur, s = mean duration of frost-free period in days. Some values are mising, where no data are available for the stations concerned (h—j are only given for diurnal types of climate).

In a climatic diagram (see Fig. 7) the months of the year are represented along the horizontal axis, from January to December for the Northern Hemisphere and from July to June for the Southern Hemisphere, so that the warm season is always in the middle of the diagram. One division on the vertical axis is equivalent to 10° C or 20 mm precipitation. The curves give the mean monthly values of temperatureand precipitation, and the scale ratio 10° C = 20 mm rain i. e., 1 : 2 holds for all diagrams. The temperature curve (t) in relation to the precipitation curve (N) is used instead of a potential evaporation curve, for which measured values are available only from a very few stations.[1] In this way the occurrence of a relatively dry season (t-curve lies below N-curve) can be depicted. In some cases a second precipitation curve is plotted in the ratio 1 : 3 i. e., 10° C = 30 mm ppt so that not only *extreme droughts* can be shown, but also less extreme *dry seasons*. Further symbols and values are given in Fig. 7.

Climatic diagrams give information concerning the mean temperature and precipitation at a particular locality (or station) over the course of the year. They also show the occurrence, length, and intensity of relatively humid and relatively arid seasons, the duration and severity of a cold winter, and the possibility of late or early frosts. With such information available it is possible for us to consider the climate from an ecological standpoint.

The arid season shown in the climatic diagram is arid only relative to the humid season of the particular climate under consideration. This is because the potential evaporation curve and the temperature curve which is used in its place are not identical, but run only more or less parallel with one another. The more arid the climate, the larger the quantitative deviation of the temperature curve below the potential evaporation curve. In absolute terms, this means that the more arid the climate as a whole, the drier is the arid season in the climatic diagram. An arid season on the climatic diagram for a station in the steppe region, for example, is not so extreme as one for a Mediterranean station or for the Sahara. This is ecologically very fitting since the drier the climate in which plants live, the more resistant they become to dryness: For the species growing in a tropical rain forest a month with less than 100 mm rain is already relatively dry, and

---

[1] The calculated evapotranspiration curves are not always reliable.

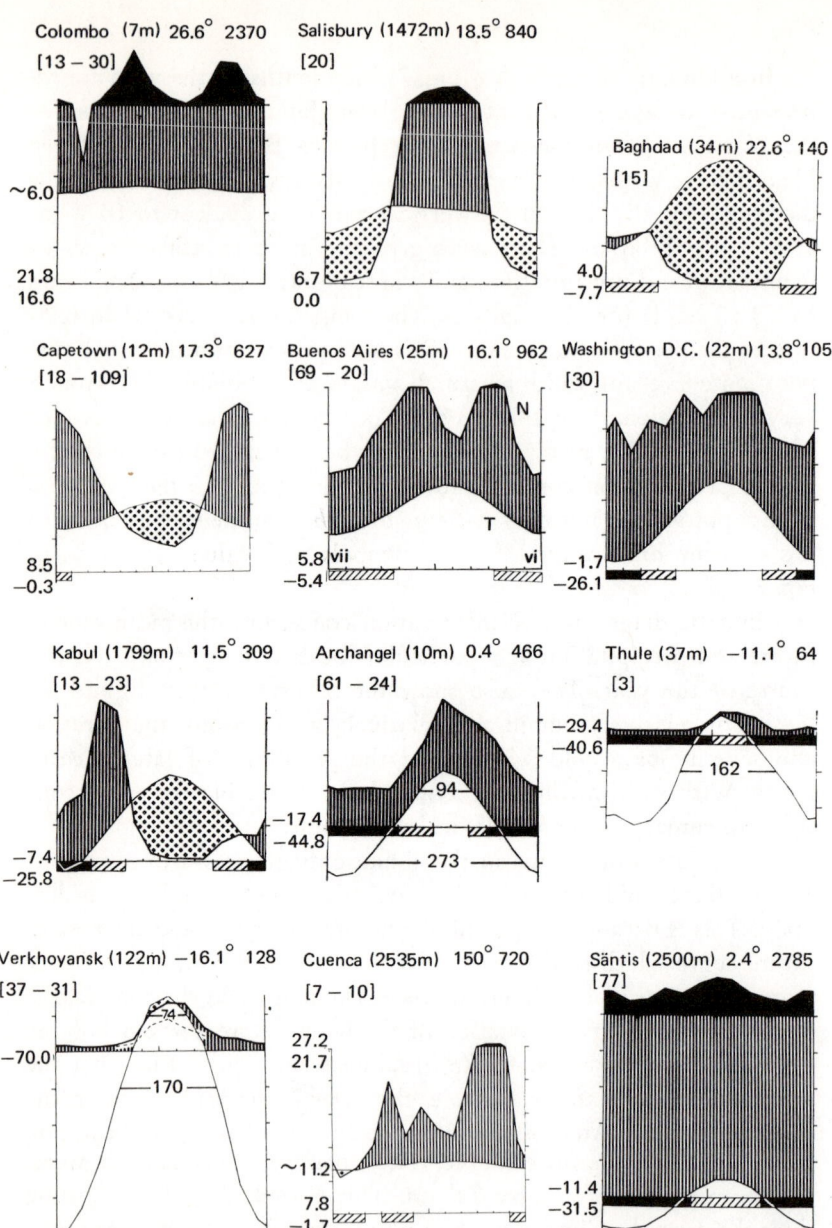

Fig. 8. Typical climatic diagrams for the climatic zones I—X. I Colombo, II Salisbury, III Baghdad, IV Cape Town, V Buenos Aires, VI Washington (see also Fig. 7 Hohenheim near Stuttgart), VII Kabul (see also Fig. 7, Ankara and Odessa), VIII Archangel, IX Thule (Iceland), VIII dry (IX) Verchojansk (Siberia), X (I—II) Cuenca in Ecuador, X (V) Säntis (Alps).

the "xerophytes" of Central Europe would rather be termed "hygrophytes" in a desert region. In order to demonstrate the way in which information can be obtained from climatic diagrams, Fig. 8 shows one diagram from each of the main climatic zones I—IX. Transitional types receive a double notation such as II (III). Climatic diagrams for all mountainous regions have the denotation X, but, in addition, the number of the climatic zone from which the mountains rise, for example, X (IV) for the Cedres diagram from the mountains of Lebanon and X (I) for that of a mountain range in the equatorial zone.

In discussing vegetational regions the appropriate diagram will always be given, since this replaces the use of long tables. The denotation of the individual vegetational zones is identical with that of the corresponding climatic zone (I—IX).

The World Atlas of Climatic Diagrams is especially useful for finding *homoclimes*, which are stations with similar climates. Figs. 9 and 10 show as an example the homoclimes of Karachi and Bombay, as well as various other Indian stations.

The climatic pattern of a country or of a continent stands out distinctly if the climatic diagrams of all the stations to be found on the Atlas are stuck onto the appropriate spot on a large wall map and the arid seasons are colored in red. Fig. 11 serves as a small-scale example with only a few (66) diagrams. The Atlas gives about 1 000 diagrams for Africa.

c Climate and Vegetation

Wherever it has occurred so far, the word "climate" has meant *regional climate* of the area in which a station is situated. The *local climate* of the habitat in which a plant community grows may differ considerably from the regional climate. Degree of exposure plays an important role: In northern latitudes the southern slopes are warmer and the northern cooler, whereas the reverse is true in the Southern Hemisphere. In the equatorial zone eastern slopes are exposed to the sun for the entire forenoon, and the western slopes are often rainy in the afternoon, whereas between northern and southern slopes there is scarcely any difference. Deep ravines have a moist climate. Since plants obtain their water from the soil and not directly from the precipitation, the situation in a habitat with a high ground-water table

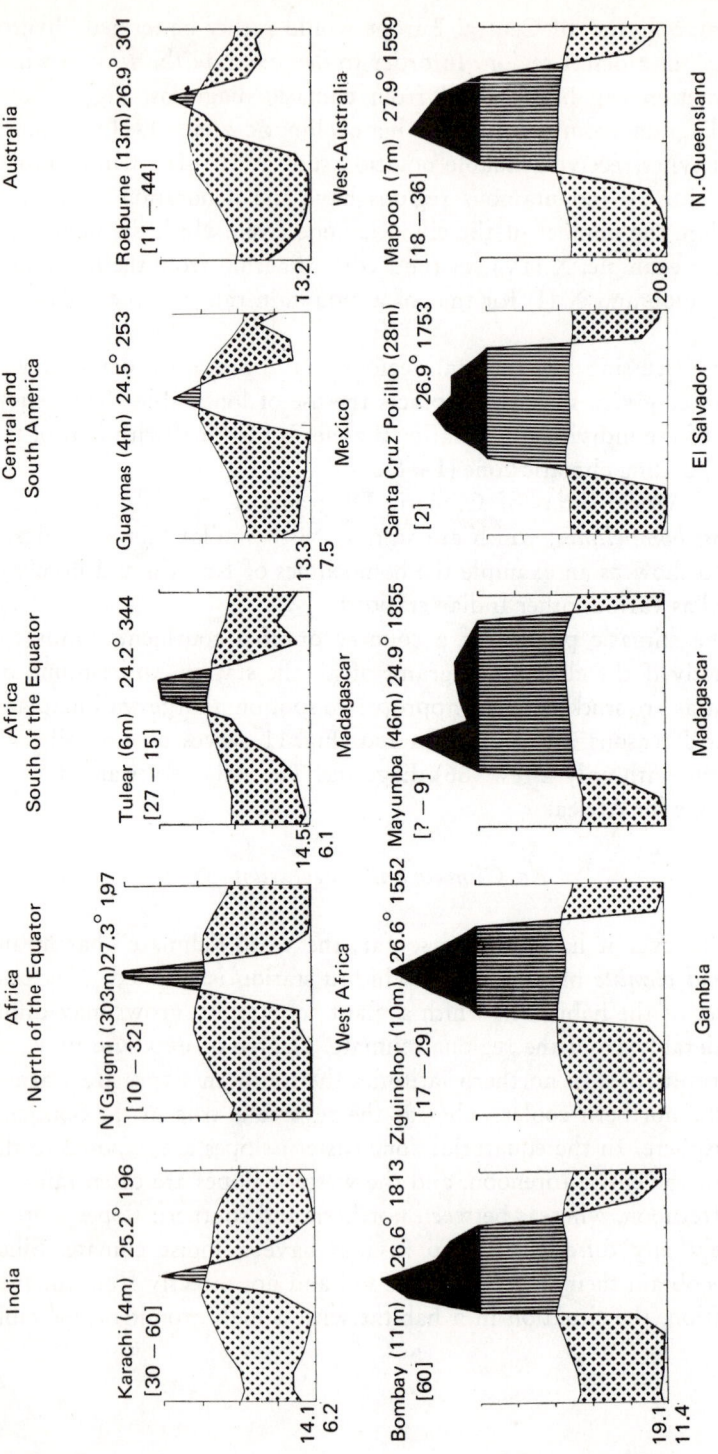

Fig. 9. Homoclimes, on other continents, of the Indian stations Karachi and Bombay.

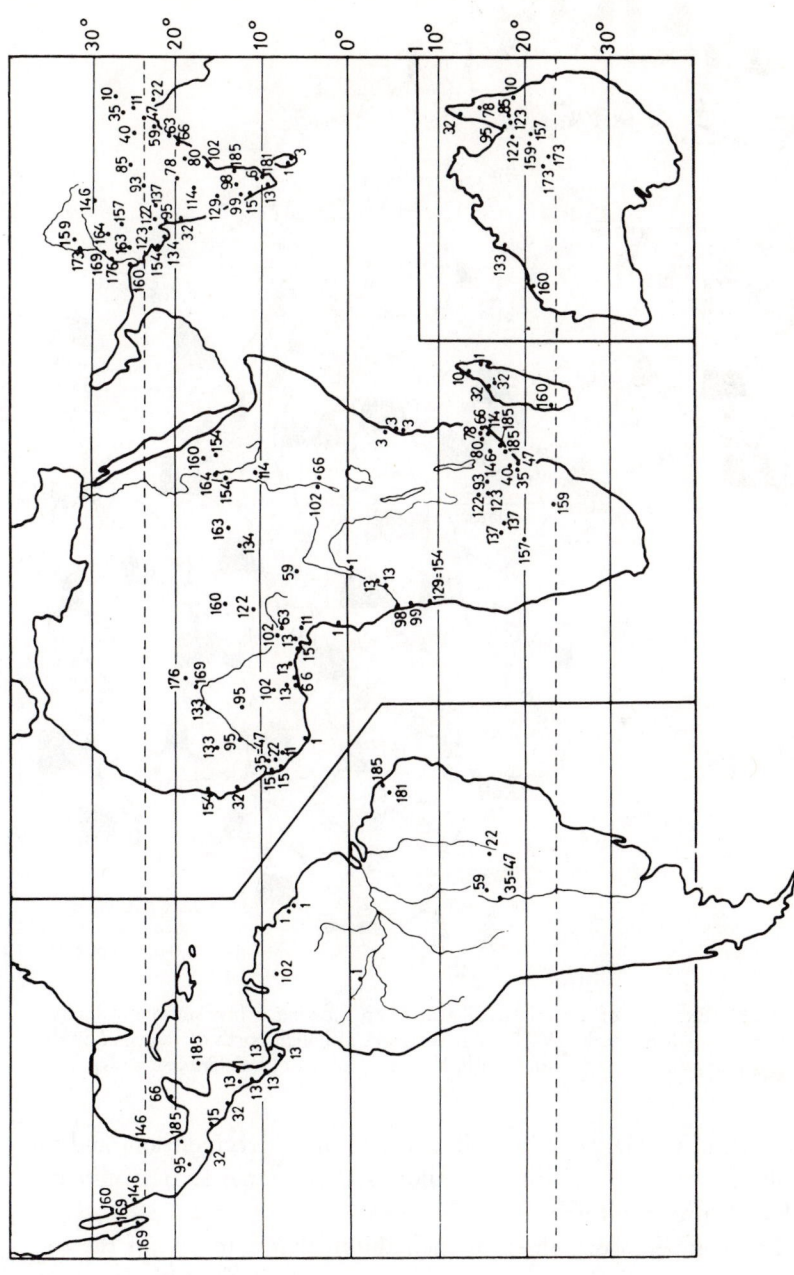

Fig. 10. The Indian stations bear the same number as that used in the Climatic diagram-World-Atlas. The same number indicates their homoclimes in other parts of the world (from Walter, a UNESCO project).

25

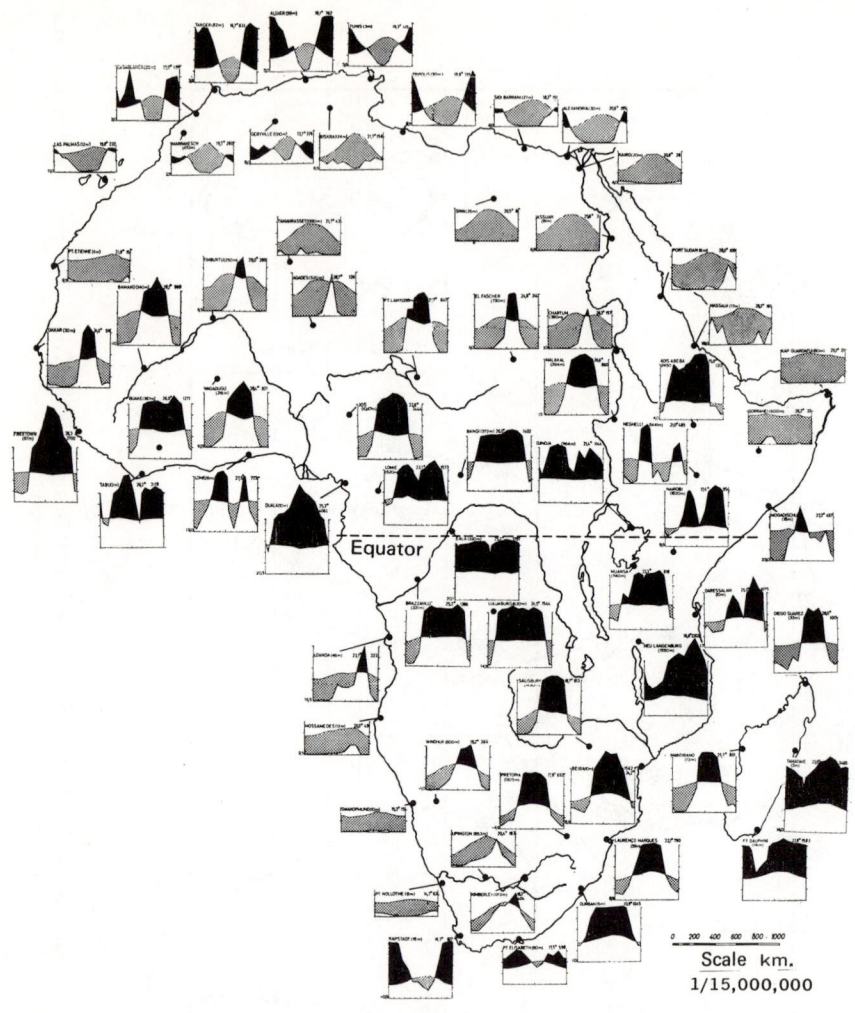

Scale km.
1/15,000,000

Fig. 11. Example of a climatic diagram-map showing a few stations. Climatic zones from north to south: IV-III-II-I-II-III-IV, although north of the equator the east is too dry (monsoon), to the south, however, too wet (S. E. Trade Winds).

is the same for plants as in a climate with high precipitation, and in a habitat with porous soil or shallow soil the same as in a climate with little precipitation.

This compensatory effect of the habitat, that is to say, of the environmental conditions, in certain biotopes makes it possible for some plants to grow outside their preferred climatic zone. The less favorable the regional climate for them, the more closely are they dependent upon a favorable habitat. This leads to the *ecological law of "Relative habitat constancy and changing biotope"*:

"If the climate within the distributional area of a plant or plant community alters in a certain direction, then a change of biotope occurs in an attempt to compensate for this climatic alteration by a change in local environment. In this way the habitat conditions in the various biotopes remain relatively constant."

Climatic factors, as measured by the meteorologists, therefore influence the vegetation only if they remain unmodified by soil or relief, as in flat areas with a low ground-water table where rainwater does not run off, or percolate too quickly through the soil, or produce water-logging. Therefore, the soils considered should be neither too heavy nor too light, but of an intermediate type such as loamy sand, where the soil type corresponding to the climate can be recognized. Such an area will be termed a climatope and the natural vegetation growing on it and corresponding to the climatic zone is the *zonal vegetation*. This vegetation can also occur as *extrazonal vegetation* in other climatic zones, in biotopes which are especially favorable to it. The zonal forest vegetation, for example, occurs extrazonally in the steppe zone along the watercourses (gallery forests, tugai). On the other hand, a steppe vegetation is to be found extrazonally in the forest zone on dry southern slopes or on porous limestone soil, and in the desert zone on cool, moist northern slopes.

Finally, there is an *azonal vegetation*, for which edaphic (soil) factors are more important than the climate, and which occurs in slightly varying composition in several climatic zones. Examples are provided by the vegetation of lakes, saline soils, cliffs, talus, dunes, and so on.

6 Vegetational Zones and Altitudinal Belts

In order to study the vegetation of the entire Earth, it is essential to begin with the largest units of vegetation, the vegetational zones, and to consider, in other words, the zonal vegetation of the individual climatic zones.

The existence of asymmetric conditions in the climatic zones of the Northern and Southern Hemispheres has already been mentioned,

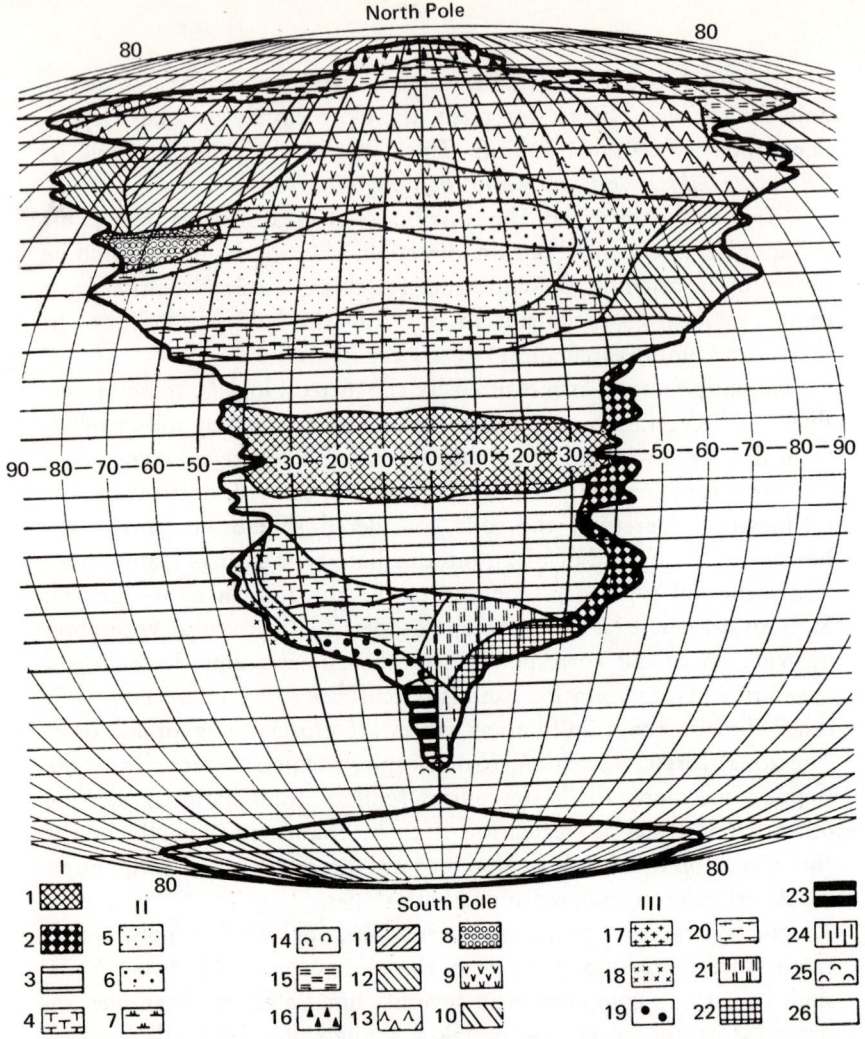

Fig. 12. "Average continent" showing the asymmetry in vegetational zones in the Northern and Southern Hemispheres (Modified from C. Troll). I. Tropical zones: 1. Equatorial rainforest, 2. Tropical rainforest with trade-wind, orographic rain, 3. Tropical deciduous forest (and moist savannas), 4. Tropical thornbush (and dry savannas). II Extratropical zones of the Northern Hemisphere: 5 hot desert, 6 cold inland desert, 7 semidesert or steppe and 8 sclerophyllous woodland with winter rain, 9 steppe with cold winters, 10 warm temperate evergreen forest, 11 deciduous forests, 12 oceanic forests, 13 boreal coniferous forests, 14 subarctic birch forests, 15 tundras, 16 cold deserts. III Extratropical zones of the Southern Hemisphere. 17 coastal and 18 mist-deserts, 19 sclerophyllous woodlands with winter rain, 20 semidesert, 21 subtropical grasslands, 22 warm temperate rainforests, 23 cold temperate forests, 24 semideserts with cushion plants, or steppes, 25 subantarctic tussock grassland, 26 inland ice of the Antarctic.

and this asymmetry is reflected in the vegetational zones. If all the continents are lumped together on a vegetational map without altering their latitudes, then an "average continent" is produced, as shown in Fig. 12. The following trends can be observed: In tropical regions and in the boreal and arctic zones, which have no equivalent in the Southern Hemisphere, the vegetational zones run roughly parallel to the lines of latitude. However, on the eastern flank, between 40° N and S the humidifying influence of the trade winds makes itself felt to such an extent that dry regions are entirely lacking. On the western side the situation is more complicated. In the subtropical zone the deserts extend to the coast, and in the Southern Hemisphere they are even confined to the coastal regions. A reversal of this situation is found, however, in the region of the cyclonic rains in latitudes above 35°. It is definitely wetter on the western side and the influence of the oceanic climate is noticeable far into the interior. The trends are easier to pick out on the "average continent" than on the map of the world showing the vegetational zones (Fig. 13), which can be interpreted as follows:

I. *The tropical evergreen rain forest zone* can be recognized in broad outline extending, in South America, across Guayana and the Amazon basin, up the eastern slopes of the Andes, including the unmapped, scattered savannas. The wet evergreen forests extend from the eastern side of Central America as far as S. Mexico, and on the east coast of Brazil as far as the Tropic of Capricorn.

In Africa the tropical rain forests are confined to the Guinea coast, the Congo basin, and the eastern part of Madagascar. In Asia they are found in the monsoon area as far as the southern slopes of the Himalayas, in Malaya, Indonesia, the Philippines, and New Guinea. They form a narrow strip on the east coast of Australia reaching southward beyond the Tropic of Capricorn.

II. Following on the rain forest zone are the *tropical moist and dry deciduous forests* and the savannas. The latter are largely anthropogenic or of edaphic origin. They predominate to such an extent in Africa that geographers usually speak of the moist and dry savanna zones. In both hemispheres this entire region/comprises the tropical climatic zone with summer rain. The larger part of Australia belongs to this vegetational zone.

III. *The subtropical deserts and semideserts* occupy only a small area in America and are confined to the southwest part of North

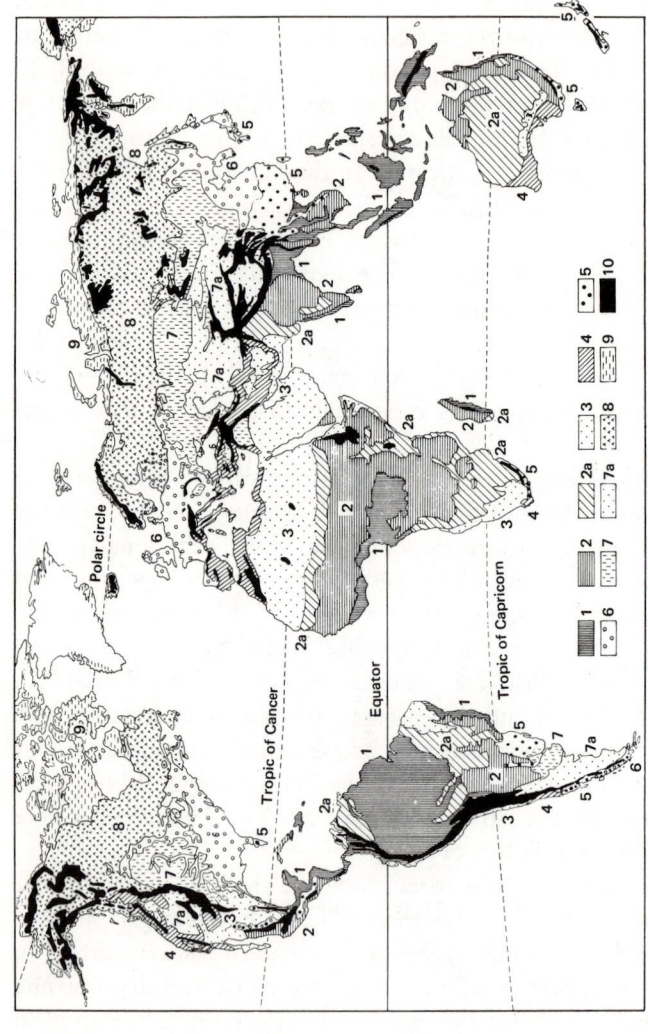

Fig. 13. Vegetational zones (much simplified, without edaphically or anthropogenically influenced vegetational regions). I. Tropical and subtropical zones: 1 Evergreen, rainforests of the lowlands and mountain-sides (cloud forests), 2 semi-evergreen and deciduous forests, 2 a dry woodlands, natural savannas or grassland, 3 hot semi-deserts and deserts, polewards up to latitude of 35° (see also 7 a). II. Temperate and Arctic zones: 4 Sclerophyllous woodlands with winter rain, 5 moist warm temperature woodlands, 6 deciduous (nemoral) forests, 7 steppes of the temperate zone, 7 a semi-deserts and deserts with cold winters, 8 boreal coniferous zone, 9 tundra, 10 mountains.

America and the narrow coastal strip on the western slopes of the Andes in Peru and northern Chile. The largest desert region begins on the Atlantic coast of north Africa, includes the Sahara and the Libyan deserts, continues in Asia as the Arabian desert, and stretches from southern Iran to India. In South Africa the deserts are confined to the southwest (Namib, Namaland, Karroo). The Kalahari, in fact, is not a desert at all. Australia has only a small region in the south with a rainfall of less than 200 mm, and, apart from this, deserts resulting from climatic conditions are entirely absent.

IV. *The sclerophyllous forests of the winter-rain regions* occupy the largest areas of the Mediterranean coasts and extend, in the mountainous regions, as far as Afghanistan. They occur in the Americas in central and southern California and in central Chile, in South Africa in the southwest corner of the Cape, and in southwestern and southern Australia.

V. *The warm-temperate, wet-evergreen forests* are at their most extensive in eastern Asia. Apart from this they are found on the southern east coast of Australia, on the North Island of New Zealand, on the east coast of South Africa, in southeast Brazil as far as northeast Argentine, in parts of southern Chile, in certain of the higher regions of Central America and Mexico, as well as on the southeastern coastal plain of North America and in Florida.

VI. *The deciduous forests of the temperate zone* occupy large areas of eastern North America, western and Central Europe, and east Asia. A very small area in Chile is all that is to be found in the Southern Hemisphere.

VII. *The winter-cold steppes and deserts* of Eurasia extend from the Black Sea almost as far as the Yellow Sea, and in the near Orient they border directly upon the subtropical desert zone. Similar vegetation in North America is found in the rain shadow occupied by the Great Basin, northward to interior British Columbia, eastward to the Rockies, and as grassland from Saskatchewan to Texas in the center of the continent. They are represented in the Southern Hemisphere by the pampa of eastern Argentina, the Patagonian semidesert, and the tussock grasslands of Otago on the South Island of New Zealand.

VIII. *The boreal coniferous zone* forms a broad belt which stretches across the entire northern part of North America and Eurasia, but is completely lacking in the Southern Hemisphere.

IX. *The tundra zone* encircles the pole in the arctic climatic region. A corresponding antarctic vegetation in the Southern Hemisphere is limited to the southernmost tip of South America and the many small Antarctic islands.

Apart from this horizontal zonation of vegetation, it is obvious that there must also be a vertical pattern of distribution, and the individual steps in this pattern are termed "altitudinal belts". The opinion that the ascending altitudinal belts are a small-scale repetition of the vegetational zones from south to north is an unfounded generalization based on the situation existing in Central Europe. Even there it is not strictly true. The order of the altitudinal belts on the northern slopes of the Alps is as follows: oak forest belt; beech forest belt; spruce forest belt; alpine belt. Moving away from the margins of the Alps to the north, via Sweden, the vegetational zones take the following order: beech forest; oak-coniferous forest; spruce forest; birch forest; tundra.

A slight similarity between altitudinal belts and vegetational zones is due only to the fact that the temperature decreases both with altitude and toward the poles, and that the duration of the vegetational period is reduced in both cases. Apart from this, a mountainous climate is very different from the climate of the higher latitudes (chapter X). This is very obvious if the seasonless alpine climate of mountains in the tropics, which have an almost uniform temperature throughout the year, is compared with the climate of the arctic tundra, with its short summers with no nights, and with the dark arctic winter.

The mountainous regions of the various vegetational zones differ so greatly in the sequence of their altitudinal belts that they will be considered individually after the vegetation of the corresponding lowlands has been dealt with.

I Evergreen Tropical Rain Forest Zone

1 Types of Vegetation of the Equatorial Zone

The mean daily temperature in the equatorial zone is the same throughout the entire year, and the length of day varies by less than

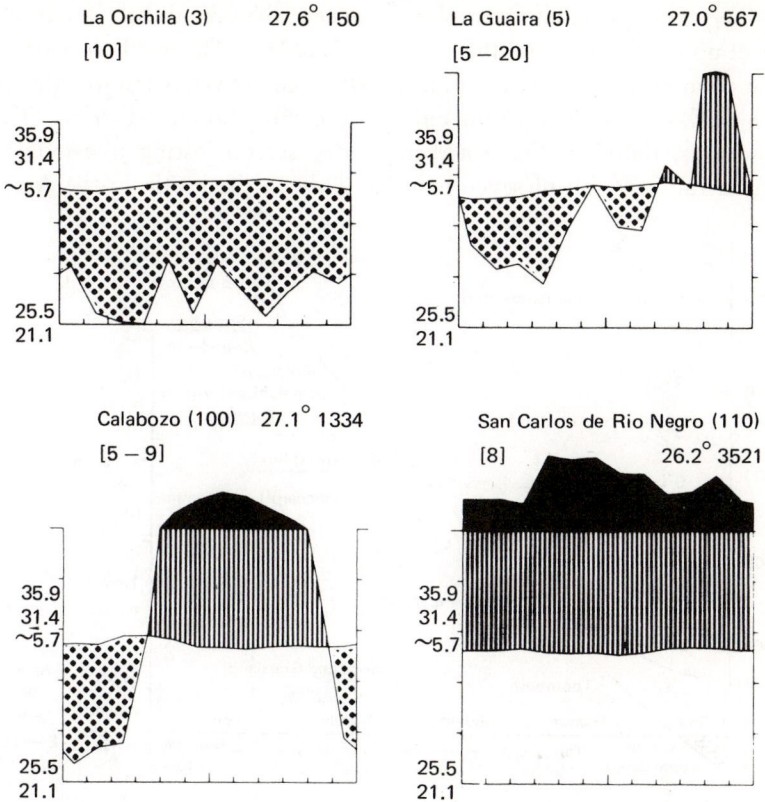

Fig. 14. Climatic diagram of a north-south profile through Venezuela (from Walter and Medina). 1 Off-shore islands, 2 Coastal station, 3 typical trade wind climate (rainy season 7 months), 4 constantly wet climate of Amazon basin.

one hour. The seasons can, in fact, only be distinguished on the basis of the distribution of rainfall. Theoretically, two rainfall maxima should be distinguishable and there should be no definite dry periods; but as a result of the influence of the trade winds and the monsoons and the presence of windward and leeward slopes in the mountains enormous variations both in quantity and distribution of annual rainfall occur. This, in turn, means that the most varied types of vegetation, from semi desert to extreme rain forest, are represented. Such large climatic contrasts can be observed distinctly within a very small area in Venezuela (Fig. 14). Venezuela lies between the equator and

33

12° N, and provides examples of every altitudinal belt from sea level up to the glaciated Pico Bolivar (5,007 m). The northern part of the country is exposed to trade winds from November until March, which, however, bring rain only to the mountainous districts. The lowlands, therefore, have a distinct dry season lasting five months and a rainy season of seven months. Only in the south, in the Ama-

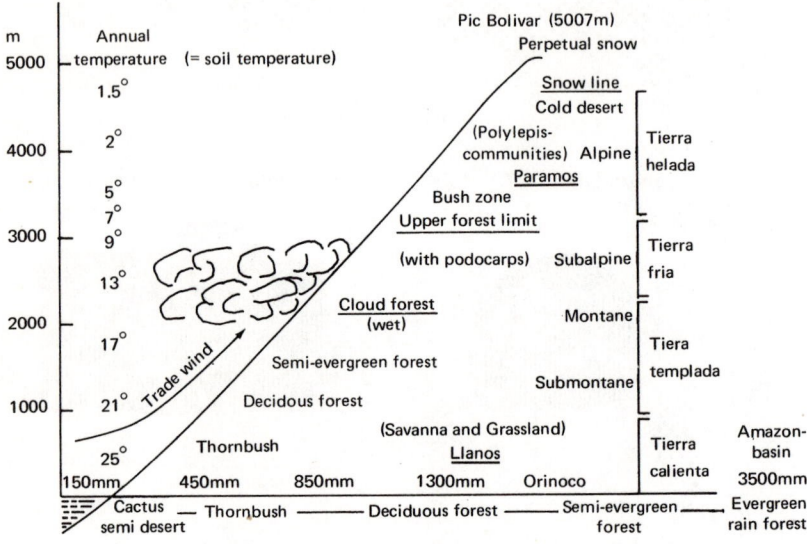

Fig. 15. Schematic representation of vegetational zones in Venezuela, from north to south, with annual temperatures in ° C (left).

zon basin, does no month have rainfall less than 200 mm. In Venezuela the annual rainfall rises steadily towards the south, from 150 mm on the island of La Orchila to more than 3 500 mm. In mountainous regions precipitation increases rapidly on the windward side up to cloud level, but decreases again above this. At the same time the mean temperature falls by 0.57° C per 100 m increase in altitude. The inner valleys of the Andes, situated in rain shadow, are extremely dry. Fig. 15 shows schematically the variations in vegetation from north to south with increasing rainfall, as well as the altitudinal belts. In the driest areas there is a cactus semidesert; these succulents store so much water that they can easily survive a dry period of six months or more (Fig. 16). With a slight increase in rainfall thornbushes and ground bromeliads occur and impenetrable thickets similar to the Brazilian caatinga are found. At an annual rainfall of 500 mm, umbrella-like thornbushes predominate *(Prosopis, Acacia)*, and are accompanied by *Bursera, Guaiacum, Capparis,* and *Croton* spp, as well as by *Agave* and *Fourcroya.*

Fig. 16. Cactus-thornbush-semi-desert with *Cereus jamaparu,* between Barquisimeto and Copeyal (Venezuela) in February (dry season).

The tree cactus *Peireskia guamacho,* which has normal leaves and is considered to be the ancestral form of all cacti, grows here. These cactus semideserts are used only as grazing land for goats.

As the rainfall increases so, too, does the number of arboreal species until true deciduous forests, extremely rich is species, commence. The tree layer is 10—20 m high and only the Bombacaceae (*Ceiba,* etc.) with their thick, water-storing trunks and the lovely flowering *Eryth-*

35

rina spp extend above this level. During the dry season such a forest presents the appearance of a European or eastern American deciduous forest in winter, although a few of the trees are already beginning to bloom at this season. A distinction must be made between dry and wet tropical deciduous forests. The latter, with a rainfall of up to 2 000 mm, attains a height of over 25 m and contains valuable timber species such as *Swietenia* (mahogany) and *Cedrela*.

The deciduous forests are sometimes cleared for coffee plantations under shade-giving trees, and for the cultivation of maize, sugar cane, pineapple, etc. Pastureland can also be produced by sowing *Panicum maximum*. Although the forests are poor in lianes they are rich in epiphytes (drought-resistant ferns, cacti, bromeliads, and orchids).

In the regions with even more rain and a still shorter dry season semi-evergreen forests occur, in which only the lower bush- and tree-layers consist of evergreen species. The evergreen tropical forest then commences and in the rainiest region of all, with no dry season, it is known as rain forest. A peculiar situation exists in Venezuela in the llanos region of the Orinoco basin, which extends far into Colombia. Instead of the deciduous forests to be expected in this climatic zone, grasslands suddenly occur, dotted with small woods or solitary trees. These are, in fact, true grasslands or savannas. Although the grass-land, nowadays used for grazing, is regularly burned, fire cannot be regarded as the primary reason for the absence of forests. The pe-culiar soil conditions prevailing in this area will be discussed later (p. 72). The following vegetational formations in Venezuela are the result of edaphic or orographic conditions and are not climatic in origin: The mangrove swamps on the sea coast and in the estuary regions, the beach and dune vegetation, the freshwater swamps and the aquatic plant communities as well as the flood-plain forests and the vegetation on dry, shallow, rocky soil. Apart from this there are the various altitudinal belts in mountainous regions.

When the trade winds encounter a mountain range crossing their path condensation takes place because of the cooling of the air masses which have been forced to ascend, and clouds and rain result. Because the force of the trade winds lets up in the late evening, the nights and early hours of the morning are clear, but for the rest of the time a layer of cloud is present at a certain altitude so that this altitudinal belt is shrouded in mist during the day. This means that, apart from the orographic rain, the trees are also loaded with drops of water from condensing mist. Because of the saturation of the atmosphere with water vapor, transpiration is entirely lacking. This extremely damp climate, which is cooler, too, on account of the altitude, favors the development of hygrophilic tropical mist forests, which are typical of all wind-exposed tropical mountain regions. The sequence of the altitudinal belts is determined by the increasing rainfall, whereas the decreasing temperature has no marked effect under 2 000 m. In Venezuela the following altitudinal belts, in ascending order, are to be found:

Cactus semidesert, thornbush, deciduous forests, semi-evergreen forests, mist forests, high montane forests with abundant *Podocarpus*, upper forest limit, alpine belt (paramos, cold desert, permanent snow).

The constantly dripping, cool mist forests differ from the hot tropical rain forests in the large number of tree ferns and in the epiphytic mosses hanging from every branch, as well as the filmy ferns (Hymenophyllaceae) covering every branch and tree trunk. In the less humid high-montane forests, often found above cloud level, epiphytic lichens predominate. The peculiarities of the upper forest limit and of the alpine belt will be dealt with later (pp. 53—59).

It is clear that quite different types of vegetation can occur within the equatorial climatic zone, and in Venezuela they are squeezed together in a very small area. The true zonal vegetation, however, is the wet evergreen forest; the other types of vegetation in Venezuela are extrazonal and the result of peculiarities in wind conditions and mountain structure (see Chapter II, p. 60 ff.).

2 Evergreen Tropical Rainforest

a Climate and Microclimate

In this wettest of all vegetational zones a month with less than 100 mm rain is considered to be relatively dry. Only in Malaya and Indonesia are there large areas that are always wet, whereas in the Amazon basin there is merely a small area on the Rio Negro. There is usually a short dry season in the Congo basin, but in India (Fig. 17) there is always one. Bogor (Buitenzorg) in Java has an extreme rain forest climate with mean monthly temperatures varying only between 24.3° C in February and 25.3° C in October: The annual rainfall is

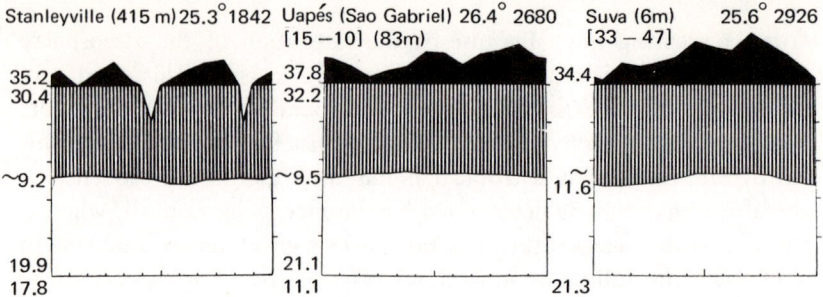

Stanleyville (415 m) 25.3° 1842   Uapés (Sao Gabriel) 26.4° 2680   Suva (6m)   25.6° 2926
                                    [15 −10] (83m)   [33 − 47]

35.2   37.8   34.4
30.4   32.2

~9.2   ~9.5   11.6

19.9   21.1   21.3
17.8   11.1

Fig. 17. Climatic diagrams of stations in tropical rainforest regions: Congo, Amazon basin, New Guinea. See also Fig. 7 (Douala, Cameroons) and Fig. 8 (Colombo in Ceylon).

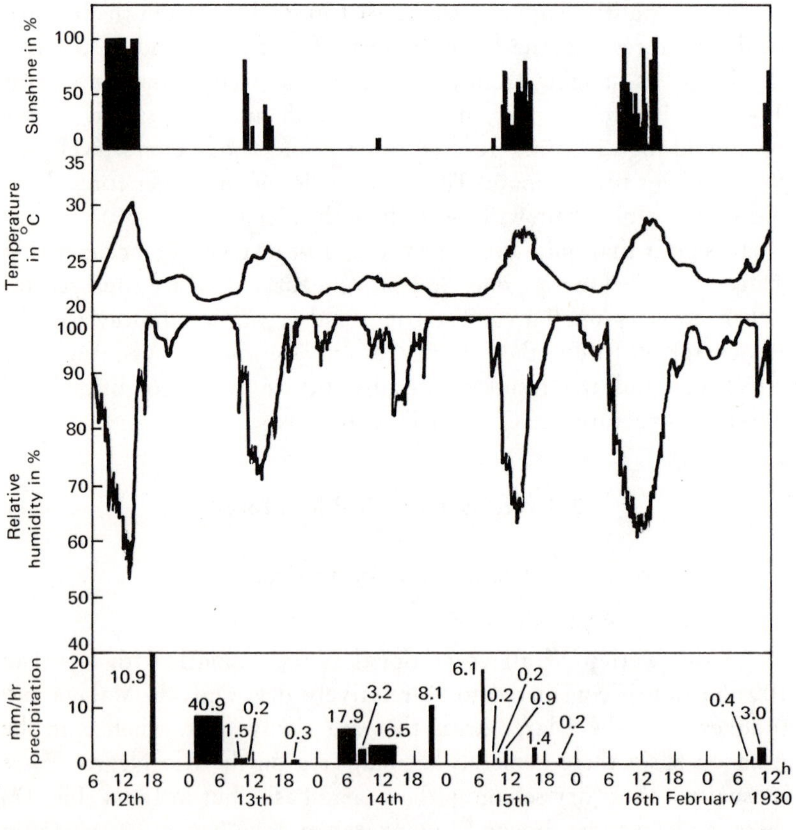

Fig. 18. Diurnal sequence of climatic factors in Bogor (Java) during the rainy season (of sunny Feb. 12th versus cloudy Feb. 14th). Figures for rain are absolute quantities of rain in mm (from Stocker).

4,370 mm with 450 mm rain falling in the rainiest month and 230 mm in the driest. Although the daily temperature variations are negligible on cloudy days (about 2 °C) they can be as much as 9 °C on sunny days, the humidity of the air varying accordingly (Fig. 18). The rain

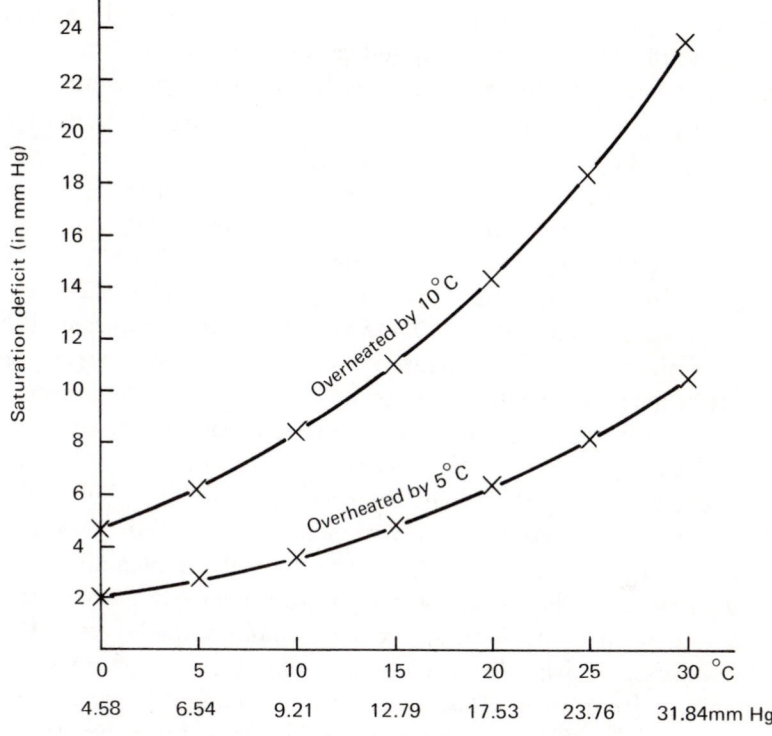

Fig. 19. Curves showing the saturation deficit at leaf surfaces if over-heating reaches 5° C or 10° C relative to the air temperature in water-vapour saturated air (lower line of figures = saturation pressure).

usually falls at midday in the form of short downpours, after which the sun shines again. When the sun is at zenith its radiation is extremely strong, which means that the leaves directly exposed to its rays heat up several degrees (up to 10 °) above the already high air temperature. Consequently, even in water-vapor saturated air large saturation deficits occur at the surface of the leaves (Fig. 19). Overheating of as much as 10 to 15° has been recorded from unshaded

Coffea leaves on clear days in Kenya. Even in the rainy season there are clear days in Bogor (Buitenzorg) when the humidity of the air sinks to almost 50 percent and the temperature rises to 30° C (Fig. 18). As a result, the saturation deficit of the over-heated leaves rises to about 40 mm Hg. Even in the wettest tropics, therefore, the leaves are at times exposed to extreme dryness for hours on end. To human beings, with a body temperature independent of the temperature of the environment, the air feels continuously oppressive and damp.

Under such circumstances it is not surprising that the leaves are adapted to a high degree to resist transpiration losses. They are equipped with a thick cuticle and are leathery or xeromorphic (for example, the rubber tree *Ficus elastica, Philodendron*); they can radically reduce transpiration by closing their stomata and thus preserve a high degree of hydrature of the protoplasm. The cell-sap concentration is usually only about 10—15 atm. It is very significant that many of these species can tolerate the dry air of heated apartments and are commonly found in Europe as indoor plants. The situation of species growing in forest shade, however, is vastly different. *The microclimate prevailing in the interior of a rainforest is much more equable,* especially on the ground itself where no direct sunlight falls. Variations in temperature are almost nonexistent at this point and the air is constantly saturated. Owing to the high air humidity even the slightest atmospheric cooling at night leads regularly to the formation of dew on the treetops and serves to moisten the leaves of the lower layers as it drips down. Light conditions, too, are of great importance to forest plants. The irregular contours of the tree canopy and the strongly reflecting, leathery leaves ensure that light penetrates deep into the interior of the forest, although its intensity at the forest floor is very small and, according to the structure of the forest, is measured as 0.5 to 1 percent of full daylight (as in temperate deciduous forests) or even as little as 0.1 percent.

b Soils and Cycling of Nutrients

Ignoring recent volcanic soils and alluvia, the soils in rain forests are usually very old, often reaching back as far as the Tertiary period. Weathering effects penetrate many meters down in silicious rock; the basic ions and silicic acid are washed out, leaving the sesquioxides

(Al_2O_3, Fe_2O_3), so that a so-called *laterization* takes place and reddish-brown loam is formed with no visible stratification into horizons. Litter decays very rapidly and wood is destroyed by termites which, because they live in subterranean colonies, are not obvious in the rain forest. The setting up of an experimental plot in the Congo was hampered by the presence of termite nests in 25 percent of the cleared ground. As a rule the reddish-brown soil lies immediately below a thin litter layer. In consequence of the high rainfall, flat areas readily become water-logged and even swampy, so that the truly typical soil is only to be found on slightly elevated ground or on slopes.

The soils are extremely poor in nutrients and are acid (pH 4.5 to 5.5), which would, at first sight, appear to be contradicted by the luxuriance of the vegetation, if it were not that *the entire nutrient reserves required by the forest are contained in the aboveground phytomass.* Each year a part of this phytomass dies off, is rapidly mineralised and the nutritional elements thereby released can immediately be taken up again by the roots. Despite the high rainfall there is no loss of nutrients due to leaching; the water in the streams has the electrical conductivity of distilled water and is at the most colored slightly brown by humus colloids. According to Went (4), mineralization of litter does not necessarily have to occur in order to provide nutrients for the plants. On very poor sandy soils in a rain forest near Manaus on the Amazon, he found that feeding rootlets of the trees at a depth of only 2 to 15 cm possess a mycorrhiza by means of which they are directly connected with the litter layer through the hyphae of the fungi. The trees can thus exploit the fungi to obtain their in organic form nutrients directly from the litter in just the same way as saprophytic species. In this manner, leaching of nutrients from the soil by rain is hindered. The quantity of leaves falling daily amounts to 4.5—12.6 g dry mass per m^2.

With such a rapid cycling of nutrients the rain forest can grow for thousands of years on the same site, but as soon as it is deforested and the wood burned an intensive leaching of the suddenly mineralized nutritional reserves occurs. Only a small portion is adsorbed by the soil colloids and can be utilized for a few years by cultivated plants. If cultivation is discontinued, a secondary forests develops which, however, never attains the luxuriant habit of the original forest. If this is once more cleared for temporary cultivation, fresh loss of nutrients takes place due to leaching until, after a series of such exploitations, the soil is capable only of supporting *Pteridium* or *Gleichenia* spp. Subsequent to the burning of such areas grasses like Alang-Alang *(Imperata cylindrica)* or other species gain a foothold.

The tropical rain forest on poor soils is inhospitable to settlers and is usually avoided by man; Very primitive tribes often seek refuge in its depths. A striking contrast is presented by the former areas of virgin forest on young, nutrient-rich volcanic soils which are today densely settled, cultivated land (Java, Central America, etc.).

The large nutritional reserves contained in the phytomass of the virgin forest depend upon its having been accumulated at a time when the rock was not so deeply weathered and the roots of the plants were still in contact with the parent rock. In totally depleted areas virgin forest can develop once more if, as a result of soil erosion, the entire soil down to the underlying rock is removed and a new primary succession is initiated. If, however, the parent rock itself is poor in nutrients, as is the case with weathered sandstone or alluvial sand, then the nutrients only suffice to support a rather weak tree- or heath-population or a sparse savanna, beneath which a *true podsol soil* can be developed. In East Borneo such a soil type has been reported, with a raw humus layer 20 cm thick (pH 2.8), an eluvial bleached horizon (pH 6.1) and an illuvial horizon (pH 5.4) below this. Podsol soils have also been found in Thailand, Indomalaya and in tropical South America. In Guiana bleached sandy soils support savanna, *Humiria* bush and *Epura* forest. This is also true of the Amazon basin in the regions drained by the Rio Negro, the waters of which appear to be black, owing to the presence of humus colloids from the raw humus soil. On similar soil on the island of Mafia (East Africa) a *Philippia* heath (Ericaceae) has been reported. If podsol soils become extremely wet they give rise to peaty soils and to tropical swamp forests which are ombrogenous bogs with dark water (pH 3). Peat thicknesses of up to 7 m have been described.

At the other extreme are the *tropical limestone soils*, associated with peculiar topographical conformations such as are found in Jamaica(5). Limestone dissolves easily in damp tropical climates, the softer parts disappear completely and the harder parts are left in the form of sharp ribs or ridges. The entire area turns into karst and a honeycomb formation develops out of the original plateau. Circular depressions or dolines, up to 150 m in depth, develop from the sink holes caused by subterranean streams. If erosion continues still further, as it has in Cuba, all that finally remains of the network of ridges are solitary towers with almost vertical faces such as the "Mogotes" or organ-pipe hills of Cuba or the "Moros" in north Venezuela. The floor of the dolines is covered with bauxitic "terra rossa" soil upon which a wet evergreen forest develops. The bare limestone rock of the honeycomb ridges presents a heterogeneous habitat, depending on whether an alkaline soil (pH 7.7) can accumulate in the few depressions or not. This explains the extremely interesting flora ranging from species of rain forest to cactus desert. In the above-mentioned areas the rainfall is less than 1000 mm. A "limestone" vegetation has not been reported from true rain forest regions.

c Structure of the Tree Stratum

The most conspicuous feature of a tropical rain forest is the large number of species constituting the tree stratum. As many as 40, or even 100, species can be counted on one hectare, most of them belonging to different families. On the other hand, there are also forests containing species of only a few families, as in Indomalaya where Dipterocarpaceae frequently dominate, and in Trinidad where the upper tree story consists of *Mora excelsa* (Leguminoseae). Large floristic differences exist between the forests of South America, Africa, and Asia, and correspondingly, the forest types are also very dissimilar, but we shall only be able to discuss the features which are more or less common to them all. Palms are almost completely absent from African rain forests although they are abundant in wet habitats in South America. The tree stratum reaches a height of 50—55 m, occasionally even 60 m, and three stories are sometimes recognizable; an upper, a middle and a lower story, although these are by no means always distinguishable. As a rule, the upper tree story is not compact, but consists of solitary giants which reach far above the other trees. It is the middle or lower stories which form a dense leaf canopy, and in such cases the trunk region is relatively free owing to lack of light and thus of undergrowth. This makes walking in the forest quite easy. However, the detailed structure of such forests varies greatly and generalizations should be treated with caution (Fig. 20). The trunks of the trees are usually slender with a thin bark; the crowns begin high up and are relatively small as a result of crowding. It is difficult to judge the age of the trees since annual rings are not present, but estimates based on rate of growth put them from 200 to 250 years. Because the soil is perpetually wet the roots do not reach far down into the soil and the giant trees achieve stability by means of enormous *plank-buttress roots* which reach, pillarlike, as far as 9 m up the

Fig. 20. Virgin forest in Siam. The liana on the trunk is *Rhaphidophora peelpla* (*Araceae*).

trunk. From the base of the trunk they radiate outward, much reduced in thickness, perhaps another 9 m. The wetter and warmer the climate, the larger the leaves on the trees, although in one and the same species the leaves that are exposed to light are always much smaller. For example, in an east African rain forest *Myrianthus arboreus* showed a ratio of 8 : 1 (largest leaf 48×19 cm, smallest leaf 16×

7 cm) and *Anthocleista orientalis* even 28 : 1 (largest leaf 162×38, smallest 22×10). Both belong to the lower tree story.

Any form of bud protection seems to be unnecessary for the trees of the rain forest although the young leaves are sometimes protected by hairs, mucus, or succulent scales, or even by specially modified accessory leaves. Even though conditions are always favorable for growth, this is periodic. The growing ends of the twigs often give the appearance of *"nodding foliage."* This is because at such a rapid rate of growth no supporting tissue is formed at first and the young leafy shoots droop downwards. They are white or bright red initially and turn green only later as they become stronger. The rapid differentiation of the leaf tips results in the formation of *"drip tips."*

A special problem is presented by the *periodicity of development* in the continuously wet tropics, where there is no annual march of temperature. As has been mentioned, there is a periodicity in growth, and this also holds true for flowering. These phenomena are, however, not confined to a particular season since external conditions are always constant. In some tree species leaf fall occurs before the new leaves begin to form (6), and a tree may even be bare for a short period. It can happen that of two adjacent trees of the same species the one is bare while the other is in leaf, or that the branches of one tree behave differently and leaf out at various times. The same is true of the flowering period; individuals of the same species may bloom at different times, or the branches of one tree may be in bloom at different times. These are all manifestations of an autonomous periodicity which is not bound to a 12-month cycle. Periods of two to four months, of nine months, and even of 32 months are observable. This means that a rain forest has no definite flowering season but that there are always some trees in bloom although, however large and beautiful the flowers, they are relatively inconspicuous against the background of predominating green.

European tree species (beech, oak, poplar, apple, pear, almond) have been transplanted to seasonless tropical mountains. The general result has been that at first the trees retained their annual periodicity of leaf-shedding, growth and flowering season. With time, however, the various branches behaved differently and finally, on one and the same tree, every stage could be found: leafless, sprouting, blooming, and fruit-bearing branches.

The tropics differ from temperate latitudes in that there is always a short 12-hour day. Temperate zone species, however, are long-day plants which bloom only when the days are long, as they are in summer. This explains their failure, as a rule, to bloom in the tropics although it is possible to substitute the effect of the long day with lower night temperatures: In Indomalaya *Primula veris* grows only vegetatively at an altitude of 1400 m, but at 2400 m it blooms and bears fruit abundantly. *Fragaria ananasa* do not bloom at low altitudes but produce many runners. In upland regions runner formation is suppressed and the plants bloom and bear fruit (7). *Pyrethrum* plantations are cultivated at 2 000—3 000 m in Kenya for their flowers, which do not develop at lower altitudes.

This endogenous rhythm of the plants adapts itself immediately to the climatic rhythm wherever there is one, as in the wet tropics, which have a short slightly drier season. On the mango tree, which is cultivated throughout the tropics, the few paler sprouting twigs on the otherwise dark crown are very obvious, but as soon as a dry season occurs the growth and flowering of all the twigs and trees adapts to it. The teak or djatti tree *(Tectona grandis)* is never bare in west Java, which is always wet, but in the dry season of east Java it loses all its leaves.

But even in the wet tropics there are such species as the orchid *Dendrobium crumenatum* which bloom on the same day over a large area. The buds in order to open need the sudden cooling down that ensues after a widespread storm. The buds of the coffee tree, too, open only after a short, dry spell, and the reproductive organs of the bamboo develop only after a dry year. In such a uniform climate certain species are extremely sensitive to even small changes in the weather.

Tropical tree species often exhibit the phenomenon of *cauliflory,* where flowers develop on older branches or on the trunk. This occurs in about 1 000 tropical species as well as in the Mediterranean *Ceratonia siliqua* and *Cercis siliquastrum.* Most of these species belong to the lower tree story and are either chiropterogamous or chiropterochorous. The fruit-eating or insect-eating bats by which their flowers are pollinated or their seeds distributed can easily reach the cauliflorous flowers and fruits.

The question of the regeneration of virgin forests in the tropics has scarcely been investigated. When a giant tree falls a large gap is left in the forest in which rapidly growing species of the secondary forest (balsa = *Ochroma lagopus* and *Cecropia* in South America, *Musanga* and *Schizolobium* in Africa, *Macaranga* in Malaya) immediately develop. They are later ousted by the upper tree story species. It has been observed that growth below individuals of the same species is often absent in the rain forest. This had led to the conclusion that such a forest is of a mosaic-like composition, each tree species being replaced by a different one. Only after a lapse of several generations can the species return to its original site so that a sort of rotation or cyclic regeneration occurs. A similar process can be observed in meadows of Central Europe, although it is not certain whether this phenomenon is generally valid for all species-rich plant communities. It would, however, provide an explanation of the observation that none of the species competing is able to attain absolute dominance and a species-rich, mixed population is the rule.

d Other Life Forms, especially Lianas and Epiphytes

About 70 percent of all species growing in a rain forest are *phanerophytes* (trees), and besides this they are quantitatively absolutely dominant. It is difficult to distinguish the shrub and herb layers from each other since herbs such as bananas and various other Scitamineae can reach a height of several meters. Even in the presence of good light conditions on the ground, undergrowth is often lacking perhaps because of competition with the roots of the trees for nitrogen and other nutrients. The lower herbs have to make do with little light and even as indoor plants in Europe they manage with very weak illumination (*Aspidistra, Chlorophytum, Saintpaulia,* the African violet).

The frequent occurrence of velvety or variegated leaves with white or red patches or a metallic shimmer deserves mention. At such high humidities guttation plays an important role and the hydrature of the protoplasm is correspondingly very high (cell-sap concentration only 4—8 atm.). The cell-sap concentration in ferns, which have the least efficient conducting system, is 8—12 atm. Heterotrophic flowering plants, saprophytes, or parasites, occur but play only a minor role, whereas lianas and epiphytes are particularly interesting types.

In dense tropical rain forests autotrophic plants have to struggle primarily for light. The higher a tree the more light its leaves receive and the greater is its production of organic matter. But to reach the light a trunk has to be formed over many years, a process involving the investment of large quantities of organic substance. Lianas and epiphytes attain favorable light conditions in a much simpler manner. Lianas do not develop a rigid stem but exploit the trees as a support for their rapidly upward-growing, flexible shoot. The epiphytes germinate in the topmost branches of the trees which thus serve them as a base.

Lianas attach themselves to the supporting tree by various means:

The scrambling lianas climb among the branches of the trees using divaricating branches armed with spines or thorns to prevent their slipping, as for example, in the climbing palm *Calamus* or the *Rubus* lianas. The root climbers put out adventitious roots to fasten themselves to cracks in the bark or encircle the trunk (many Araceae). The winding or twining plants have long, rapidly-growing, twining tips to their branches and very long internodes upon which the leaves are at first underdeveloped, while the tendril-climbing lianas possess modified leaves or side shoots which are sensitive to touch stimuli and serve as grasping organs. Light is essential for the growth of lianas and for this reason they are common in forest clearings and grow simultaneously with the trees of the forest regrowth, until they finally reach the level of the forest canopy. Tropical lianas are long-lived as compared with those of temperate latitudes. Their axial organs are equipped with secondary thickening but their stems remain pliable so they can folow the growth of the supporting trees. The woody structure is not compact but the ligneous portions are permeated by strands of parenchymatous tissue or broad medullary rays (anomalous thickening). In cross section the vessels are large and have dividing walls so that despite the small diameter of the pliable stem the crown of the liana receives a sufficient supply of water. Should the supporting tree die and decay the lianas still remain fastened to the tops of the other trees and their stems hang down like ropes. They often slip down and lie on the ground in loops around their own lower ends, but the shoot tips work their way up again. If this happens several times the stem can attain a great length: a total length of 240 m has been measured in *Calamus*. Complete deforestation is especially favorable to the development of lianas, which are, for this reason, much more numerous in secondary than in virgin forests, where they prefer the margins. Ninety percent of all liana species are confined to the

Fig. 21. Epiphytes on a tree branch in rain forest, Brazil. *Bromelia* rosette on the right, pendent 3 *Rhipsalis* spp., lanceolate leaves of *Philodendron cannaefolium*.

tropics; in western India 8 percent of all species are lianas. In Central Europe only three species of woody lianas are found: The root-climbing ivy *(Hedera helix)*, the scrambling Traveler's Joy *(Clematis vitalba)* and the twining wild grape *(Vitis sylvestris)*. Although European blackberries *(Rubus* spp.*)* do not grow high above the ground, in New Zealand they grow as thick as a man's arm and reach up to the treetops. The difficulty of water transport is probably what confines lianas mainly to the wet tropics. In a dry climate the large suction tension (low water potential) built up in the leaves causes disruption of the water column by overcoming the cohesion in the wide vessels. The European woody lianas, too, are at their most abundant in the wet flood-plain forests.

A wealth of interesting types is found among the epiphytes (Figs. 20 and 21). The fact that epiphytes germinate high up on the branches of trees provides favorable light conditions but brings with it the problem of water supply, since the constant water reservoir otherwise provided by the soil is lacking. Epiphytic habitats can be compared with rocky habitats and epiphytes do, in fact, grow well on rocks if light conditions are satisfactory. Since they can take up water only while it is actually raining, the rainfall frequency is of greater importance to them than the absolute quantity of rain. On windward mountainous slopes the frequency of rainfall is greater than on flat land on account of the ascending air masses. This is why montane forests, especially the mist forests where the leaves are constantly dripping, are rich in epiphytes. To be able to withstand the large intervals between rain showers epiphytes must either be capable of resisting desiccation without undergoing damage, as is the case with many epiphytic poikilohydric ferns, or they must store water in their organs as do the succulents of dry regions. A whole series of cacti has changed over to an epiphytic way of life *(Rhipsalis, Phyllocactus, Cereus* spp.). Epiphytes conserve their water economically just as do the succulents; many orchids possess leaf tubers as water reservoirs, and the majority of orchids, bromeliads, peperomiads and other epiphytes have succulent leaves. The velamen of the aerial roots of orchids ensures rapid uptake of water during showers; bromeliads are equipped with water-absorbing scales which take up the water collecting in the funnels formed by the leaf bases or serve to retain the water by capillary forces and then suck it in.

The roots of epiphytic bromeliads serve only as adhesive organs and are completely absent in *Tillandsia usneoides,* which appears very superficially similar to the lichen *Usnea. Myrmecodia, Hydnophytum,* and *Dischidia* spp develop special cavities, sometimes inhabited by ants. Ferns, which cannot tolerate dehydration, can produce their own soil by collecting litter and detritus between their funnel-like, erect leaves *(Asplenium nidus)* or with the aid of special overlapping "niche" leaves *(Platycerium).* In this way a soil is formed which is both rich in humus and which retains water so that the roots growing into it are well provided for. In a forest densely populated by epiphytes the epiphytic humus can amount to several tons per hectare and in this way a new biotope is created far above ground level. Dripping water and dust bring in nitrogen and other nutrients, ants colonize the area and build their nests and also drag in seeds which then germinate and grow into flowering plants. Such "flower gardens," as they occur in South America, harbor a special fauna and microflora: mosquito larvae, water insects, and protista inhabit the funnels of the bromeliads, which often achieve considerable dimensions. It should be mentioned that the insectivorous *Nepenthes* (pitcher plants) as well as certain *Utricularia* species can also grow epiphytically.

Epiphytes are distributed by means of spores (ferns), dustlike seeds (orchids) or berries (cacti, bromeliads) that are eaten by birds, the seeds reaching the branches of the trees in the droppings. Many epiphytes are able to endure a long period of drought. Among them are orchids, which lose all their leaves, densely scaled Tillandsias and poikilohydric ferns. Epiphytes are also found in the dry type of tropical forest.

Coutinho recently demonstrated the occurrence of the de Saussure effect *(Crassulacean metabolism) i*n some epiphytes in Brazil, i. e., the *nocturnal absorbtion of* CO_2 through open stomata and its binding as organic acids (carboxylation) (8). The latter are decarboxylated during the day and the carbon dioxide thus set free is immediately assimilated while the stomata are still closed. By means of this process loss of water due to daytime transpiration can be avoided, and it has also been found to take place in the succulents of dry regions.

In semiarid wooded regions there are no epiphytes, but in their place hemiparasites such as mistletoes *(Loranthaceae)* are often found on the trees. Mistletoes are even to be seen on the stem succulents on the edge of the desert. Mosses and hymenophyllous filmy ferns, on the other hand, require constant humidity and are thus the typical epiphytes of the mist forests algae and mosses that grow on leaves.

Hemiepiphytes occupy; also an intermediate position between lianes and epiphytes. Many Aroideae germinate on the ground and then grow upward as lianas, usually as root climbers. In time, the lower part of the stem dies off so that the hemiepiphytes turn into epiphytes although still remaining in contact with the ground by means of their aerial roots. Even more striking are the *"strangling" trees* of which the numerous "strangling" figs are the best known *(Ficus* spp*).* Such "stranglers" occur in many different families, as for example the *Clusia* spp. (Guttiferae) of South America, the New Zealand *Metrosideros* (Myrtaceae) and many others. They germinate as epiphytes in the fork of a branch and at first put out only a small shoot and a long root. The latter rapidly grows down the trunk of the supporting tree and enmeshes it. Only when the root reaches the ground does the shoot begin to grow; at the same time the roots thicken and prevent secondary thickening in the supporting tree. This is thereby "strangled," dies and decays. The root mesh of the "strangler" unites to form a trunk bearing a broad crown. Such trees can attain enormous dimensions and it is no longer apparent that they began life as epiphytes. Palms without secondary growth are not strangled and achieve longevity, until their leaves are too greatly overshadowed by the crown of the "strangler."

3 Altitudinal Belts of the Tropical Mountain Ranges

a Cloud Forests

Ascending air masses bring increased precipitation to mountain slopes, unless the slopes are sheltered from the wind, and even if there is a dry period in the lowlands this becomes shorter with increasing altitude, or even disappears altogether. It is therefore not surprising that the montane forests are particularly luxuriant and rich in epiphytes. Tropical mountain slopes are generally very steep so that the soil is well drained. In ascending such mountains change in the vegetation with decreasing temperature is initially scarcely noticeable. Finally, at cloud level, where a state of maximum humidity prevails, the cloud forests commence. They are not connected with a definite altitude but rather with the cloud level itself, which, in turn, is dependent upon the humidity at the foot of the mountain: The greater the humidity there, the lower the cloud level. In a climate with both a wet and a dry season the clouds are higher during the dry season. Cloud forests can be found between 1000 and 2500 m above sea level, or even higher still, and the variety of temperature conditions which can prevail accounts for the floristic differences exhibited by such forests. Even the height of the tree stratum decreases with increasing altitude, and the trees of very high forests are gnarled and stunted (elfin forest). But the common feature of all cloud forests is their profusion of epiphytes. Whereas the number of warmth-loving, epiphytic flowering plants decreases with increasing altitude, the ferns, lycopods and, above all, the filmy ferns (Hymenophyllaceae) and mosses become more abundant. The branches of the trees are typically draped with curtain-like mosses, themselves covered with water droplets, while filmy ferns, which roll up as soon as the humidity drops slightly below 100 percent, form a green covering on the trunks and branches. The forest floor is often carpeted with bright green *Selaginella* spp. and the tree ferns, which prefer a damp, cool climate, are numerous. In many tropical mountain regions the wettest altitudinal belt is characterized by palms or dense bamboo groves.

Increasing altitude also brings with it changes in the soil: The reddish-brown loams of the lower belts are gradually replaced by more yellow types; at the same time a mull horizon appears and the clay content decreases. Further up, a slight podsolization is detectable and eventually true podsols, with a leached horizon and raw humus, occur. In the perhumid cloud belt gley soils are found.

b The Upper Forest Limit

Precipitation decreases rapidly above the cloud belt, and if the forest extends further upward then the leaves of its trees become smaller and more xeromorphic. Conifers of *Podocarpus* species, which posses narrow, leaf-shaped structures in the place of needles, are present. Mosses are replaced by beard-like lichens and, finally, at the upper forest limit, a shrub zone commences. This occurs at a much lower level in the tropics than in the subtropics: An altitude of from 3,100 to 3,250 m above sea level is given for the Venezuelan Andes. The shrub zone is narrow and the shrubs themselves get smaller the higher up they grow. In Venezuela the highest are found, sheltered by rocks, at 3,600 m.

The question of which factors are responsible for determining the timber line in the tropics is still unresolved. The fact that precipitation decreases with increasing altitude would suggest that aridity sets the limit. On the other hand, it might well be that frost is the limiting factor since at this elevation temperatures can drop below freezing point. Personal investigations in Venezuela, however, suggest that in all probability it is the soil temperature that is of ultimate importance, although as always with such phenomena a variety of other factors is involved (9). In the equatorial zone, where the climatic fluctuations are of a diurnal rather than an annual nature, the temperature variations do not penetrate very deeply into the ground. On shady ground, the temperature at a depth of 30 cm is constant throughout the year and is the same as the mean annual air temperature measured by the meteorologists. Armed with a spade and a thermometer it is possible to determine the annual mean temperature at any spot in the tropics in a few minutes. In dense forests the temperature is constant immediately below the ground surface, and it is this temperature that is decisive for the root system of the plants. Although the minimum temperature requirements for root growth of tropical trees are unknown, it is, however, known that the enzymes responsible for protein synthesis in the roots have a temperature minimum well above 0° C. This means that at temperatures considerably above freezing point tropical species can be "chilled" and die. Assuming the temperature minimum of the roots of trees situated at the timber line is around 7—8° C this would coincide exactly with the temperature of the soil at the timber line in Venezuela, which is made up of typical tropical species and is completely lacking in holarctic species. If this assumption holds true it would explain the higher timber line in the subtropics. In these regions there is already an annual march of temperature and in the summer the temperature of the soil rises considerably above the mean annual temperature. Arboreal species are able to exploit this favorable season which is lacking in climatic zones where the fluctuations in temperature are of a diurnal rather than an annual nature. At an altitude of 3,864 m in the Pamir, for example, the August temperature at a depth of 1 m remains above 10° C and the maximum temperature at a depth of 40 cm exceeds 20° C. The temperature minimum of the roots of species typical of a tropical shrub zone must be lower, an explanation apparently supported by the presence of many holarctic species of the family Ericaceae and of the gener Hypericum, Ribes, etc.

c Alpine Belt

In the wet tropics the alpine belt is termed the páramo. It is described as being perpetually wet, misty, desolate, and cold, although annual temperature and precipitation curves are not given. For Venezuela such curves have been made available to the author and it can

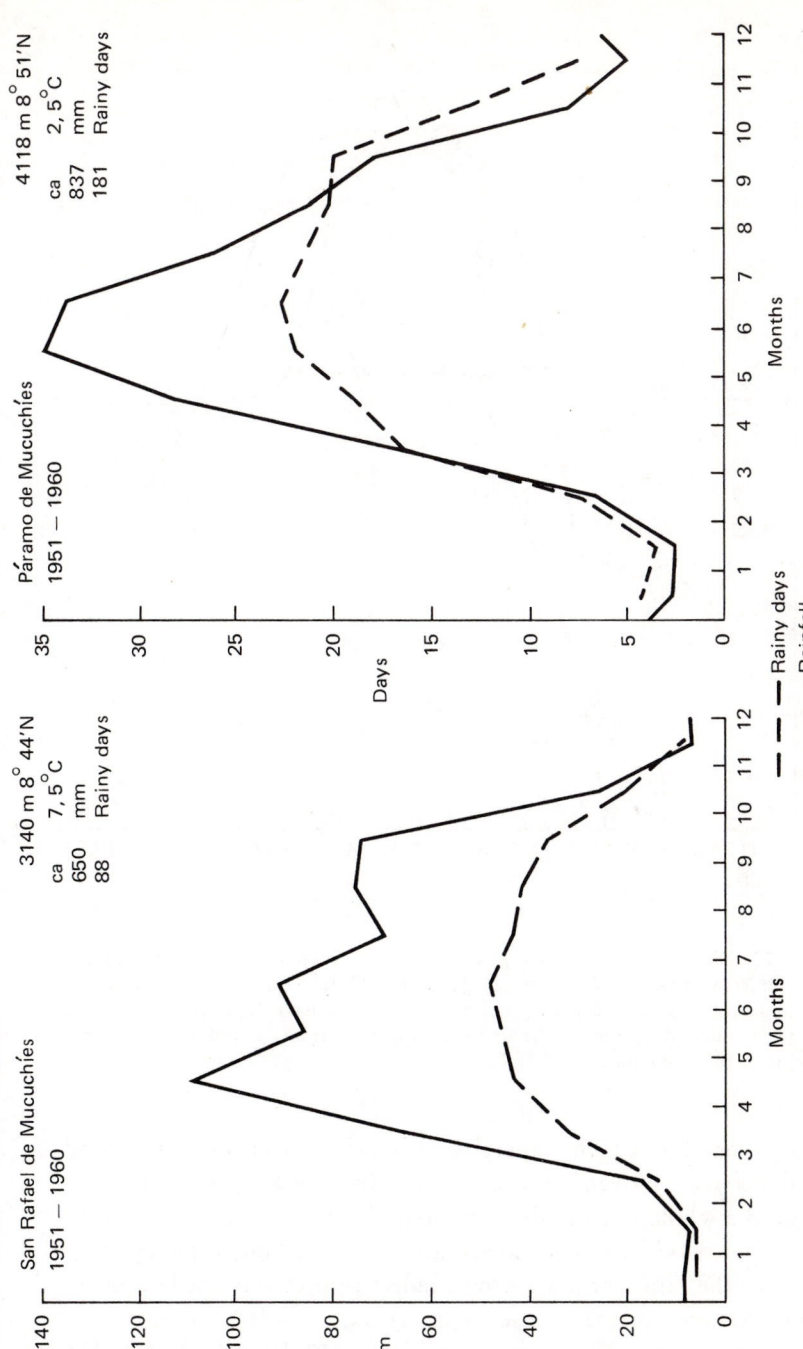

Fig. 22. Rainfall distribution in the páramo belt. Drier season from November to March.

San Rafael de Mucuchíes
1951 – 1960

3140 m 8° 44'N
ca 7,5°C
650 mm
88 Rainy days

Páramo de Mucuchíes
1951 – 1960

4118 m 8° 51'N
ca 2,5°C
837 mm
181 Rainy days

Rainy days
Rainfall

Months

Days

mm

55

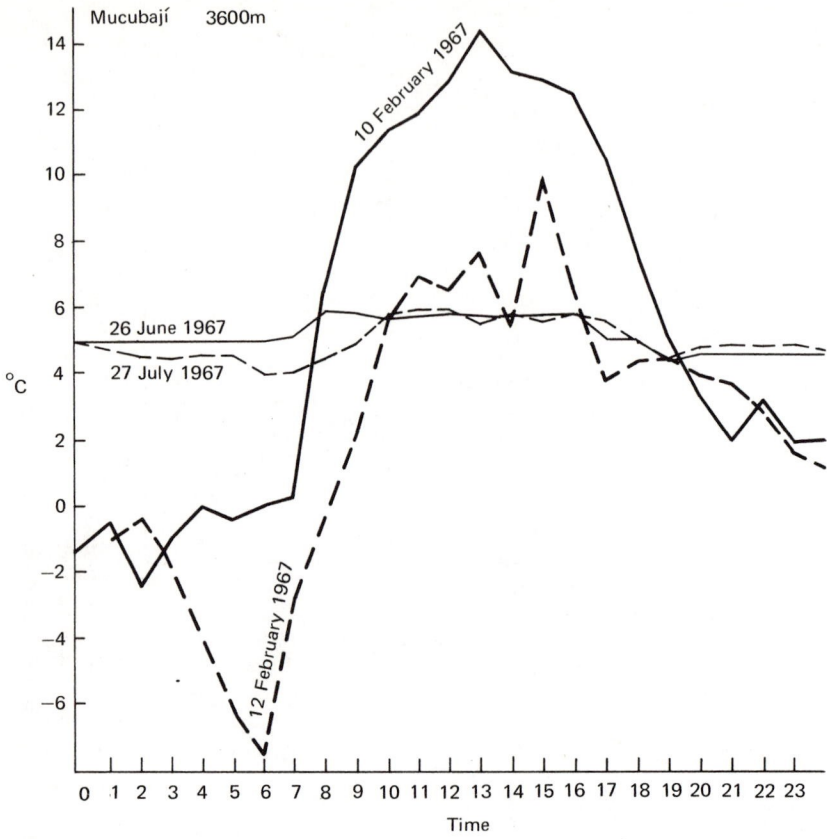

Fig. 23. Diurnal course of temperatures in the meteorological hut (páramo belt at 3600 m above sea level) on 26 June and 27 July during the rainy season (variations only 1.6 and 2.0° C resp.) and on 10 February (hottest day) and 12 February (coldest day) during the dry season (variations 17.0 and 17.5° C resp.). t-Maximum 14.5, t-minimum —7.5° C.

be seen from Fig. 22. that there is little rain during the trade-wind period between November and March. In January the author experienced a whole week in the páramo under a cloudless sky when the cloud layer was situated at lower altitude. The hourly temperature values for the rainy, or dry seasons, reflect respectively the lack of incoming radiation or the intense incoming radiation (10 February) and the high net outgoing nocturnal radiation (12 February) (Fig. 23).

The coldest day of 1967 followed the warmest almost immediately. The air at 3 600 m usually reaches a daytime temperature of 10° C during the dry season, although it freezes at night. Furthermore the plants themselves are exposed to much greater extremes than are the thermometers in the shelters. Apparently this continuous freezing and thawing does the plants no harm because it coincides with the main flowering sea-

Fig. 24. Páramo de Mucuchíes, 4280 m above sea level (Venezuela, Andes). In the foreground a cushion of *Aciachne pulvinata* in the process of disintegrating. Further back, numerous *Espeletia alba*.

son. The upper soil layers harboring the roots of the páramo plants warm up during the day at this season to temperatures above the annual mean. A rocky habitat is apparently more favorable than a wet, cold soil. On the basis of soil measurements the mean annual temperatures can be given as follows: at 3 600 m 5,0° C (agrees with meteorological data), at 3 950 m 3,9° C and in the firn or permanent snow at 4 765 m —1,5 to —3,5° C. A decrease in temperature forces the plants to spread their roots nearer the surface, and this inevitably leads to a sparser and sparser vegetation, until finally it ceases altogether some 100 m below the permanent anow line. This cold desert belt is typical of tropical mountain regions. In higher latitudes (in the Alps) the plants take advantage of a favorable season to grow in snow-free places above the snow line. The soil in the páramos remains moist even during the dry season so that the vegetation does not suffer from drought and gives a hygromorphic impression.

It would be interesting to ascertain whether there are páramos with no sunny, dry season and, if so, how the vegetation reacts to such a situation. The floristic composition of the páramos in South America, Africa and Indonesia vary greatly, each area having its own peculiarities. It is striking that, apart from the low plants which give the impression of hugging the ground, there are also tall species, mostly Compositae, with a proper stem and large, bushy, upright leaves covered with thick, white hairs (Fig. 24). These are, in the Andes, *Espeletias* (27 species), in equatorial African regions, *Senecio* tree species, and in Indonesia, species of *Anaphalis*. The woolly, candlelike form of *Lupinus* and *Lobelia* is also conspicuous. On Kilimanjaro extremely hairy species of *Helichrysum* grow as far up as 4 400 m, although whether or not their hairiness serves as a heat insulation providing protection against sudden variations in leaf temperature has not been investigated: At such altitudes the passage of a cloud on a sunny day invariably leads to a sharp drop in temperature. The upper vegetation limit lies at about 4 000—4 500 m and probably coincides with a mean annual temperature of about 1° C.

It is especially remarkable that in the South American Andes, in the middle of the alpine belt at an altitude of approximately 4 200 m and with a mean annual temperature of 2° C, small stands of *Polylepis sericea* trees (Rosaceae) occur. They are invariably found on steep east- or west-facing talus slopes, consisting of large boulders, and are exposed to the sun in the morning or afternoon. The roots of *Polylepis* sometimes go down as far as 1,5 m, and the only possible explanation for this isolated occurrence of trees 1 000 m above the timber line is that the talus slopes provide particularly favorable temperature conditions. Incoming radiation warms up the air layer nearest the ground very strongly; the heavier cold air between the boulders flows out at the lower end of the talus so that warm air is drawn in between the boulders in the upper part. Therefore the temperature conditions for the root systems of the *Polylepis* trees are more favorable than they would normally be at an altitude of 4 200 m. This explanation finds support in the observation that the lower part of the talus slope is devoid of trees and is often even completely bare (9).

4 The Evergreen Tropical Rain Forest as an Ecosystem

The most complicated of all biogeocenoses are to be found in this type of vegetation, and this, together with the lack of well-equipped laboratories in the immediate vicinity of virgin forest is responsible for the meagerness of our knowledge about productivity and energy flux in such biogeocenoses. The existence of such luxuriant vegetation would suggest a very high primary production. The preliminary estimates of about 100 t/ha dry substance, proved to be too high. It must be borne in mind that the phytomass of tropical rain forests has a very high water content (75 to 90 percent for the herbaceous parts), and that although the green leaves are able to assimilate CO_2 throughout the year nocturnal respiration losses are especially large owing to the high temperatures. Wood production in tropical forest plantations can attain values of 13 t/ha, which is only about twice that of a good European beech forest and occurs only because the vegetation period is twice as long. The leaf area index (LAI), or the ratio of total leaf area to the total ground area covered in a particular ecosystem, is of great significance for the productivity of the biogeocenose. This index has been determined for a tropical rain forest on the Ivory Coast and was found to be about the same as that for a healthy beech forest (10). This surprising result explains why light conditions on the forest floor of tropical rain forests are not much less favorable than in dense temperate deciduous forests. As was to be expected the gross production of virgin forest on the Ivory Coast proved to be very large (52,5 t/ha). However, 75 percent of the organic substance produced is lost by respiration: Respiratory losses of the leaves = 16,9 t/ha, of the axial organs = 18,5 t/ha, and of the roots (estimated) = 3,7 t/ha; in all 39,1 t/ha. Since the losses due to respiration in a beech forest amount to only 10,0 t/ha, i. e., 43 percent of the gross production of 23,5 t/ha, it is undertandable that the primary production of the tropical rain forest is no higher than that of a well-tended beech forest in Central Europe:

52,5 − 39,1 = 13,4 t/ha   Tropical rain forest
23,5 − 10,0 = 13,5 t/ha   Beech forest

These results should be compared with others, such as those from Indomalayan rain forest. In the tropical rain forest of Thailand the leaf area index was 12,3, the gross production 124 t/ha and the primary production nearly 30 t/ha, or twice as high as in a beech forest corresponding to twice as long a growing season (12 months).

An indirect method of checking on the primary production of a tropical rain forest uses determinations of soil respiration. When a biogeocenose is in a state of equilibrium then the total organic substance produced each year is remineralized. Whatever the amount reaching the soil, an equivalent quantity, plus the root respiration, has to leave the soil again as CO_2 (soil respiration). The 24-hour values for soil respiration are probably constant in regions where the climate is the same all year round. Measurements were made over the course of an entire year in a Venezuelan cloud forest at an altitude of 1 126 m with a mean annual temperature of 14° C and a rainfall of from 2 000 to 3 000 mm. Litter production amounted to 7,8 t/ha, which is somewhat higher than in the rain forests of the Ivory Coast. The quantity of dry material broken down by soil respiration varied from 3,9 to 10,1 t/ha and averaged 5,0 t/ha. This is rather smaller than the quantity of litter because litter is already partly decomposed before being incorporated into the soil. Dead wood and epiphytes are largely broken down above ground and thus play no part in soil respiration. In any case, these values also indicate that the primary production of the tropical rain forest is not exceptionally high (11).

II Vegetation of the Tropical Summer-Rain Zone

1 Changes in Vegetation Connected with Increasing Length of the Dry Season and Decreasing Rainfall

In the course of our discussion of the vegetation of Venezuela, with regard to decreasing annual precipitation and lengthening dry season, the following zonation was mentioned:
evergreen rain forest, semi-evergreen forest, deciduous forest.
This zonation is only rarely observable within the equatorial climatic zone since a stepwise increase in rainfall, such as is found in Venezuela, is exceptional. As one leaves the Tropics, however, such

zones can be distinguished. In moving away from the equator, the tropical climatic zone with zenithal summer rains is penetrated more deeply, the absolute rainfall decreases continuously, and the rainy season becomes shorter. In contrast to Venezuela, the annual march of temperature becomes more marked, and the dry season occurs at the cool time of year. But since the latter means a rest period for the vegetation, the temperature variations are of little significance from this point of view.

It has already been mentioned that, when a short dry period occurs in the very wet tropical region, the endogenous rhythm of tree species adapts itself to the climatic rhythm. The general character of the forest remains unchanged, but many trees lose their leaves at the same time, or sprout or flower simultaneously so that the vegetation does in fact exhibit definite seasonal changes in appearance.

If the dry season becomes even longer, the type of forest changes. The upper tree story is made up of deciduous species. In South America these are the large, thick-trunked Bombacaceae, whereas the lower stories are still evergreen, so that the forests can be termed tropical semi-evergreen.

With a further decrease in rainfall and a lengthening of the dry season, all of the arboreal species are deciduous so that the forest is bare for shorter or longer periods of time. These are the moist or dry deciduous tropical forests, climatic diagrams for which are shown in Fig. 25.

The question now arises as to whether the amount of rain or the duration of the dry season determines the structure of the forest. The diagram in Fig. 26 shows that both factors are ecologically important and that neither should be considered alone. The course taken by the limiting lines indicates that for the wet type of forests the duration of the dry season is more important, while the absolute annual rainfall is of greater significance for the dry types.

In Africa the above-mentioned succession of forests is difficult to find. Dense population and shifting cultivation have led to the deforestation of semi-evergreen forest and wet deciduous forest. They lend themselves better to clearing than the rain forests since they can be burned during the dry season. Besides this, the rainfall is sufficient to ensure an annual crop yield, which is not the case with the dry type of forest. Anthropogenic savannas with either tall grass (moist

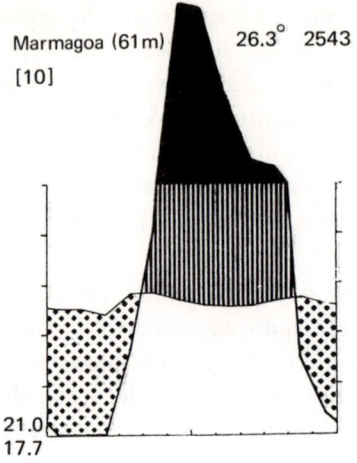

Marmagoa (61m)
[10]
26.3° 2543
21.0
17.7

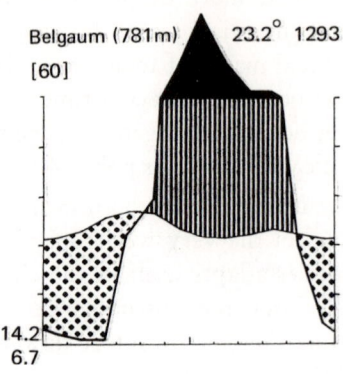

Belgaum (781m)
[60]
23.2° 1293
14.2
6.7

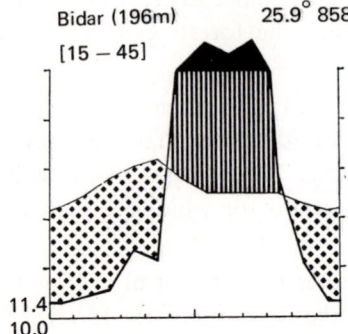

Bidar (196m)
[15 – 45]
25.9° 858
11.4
10.0

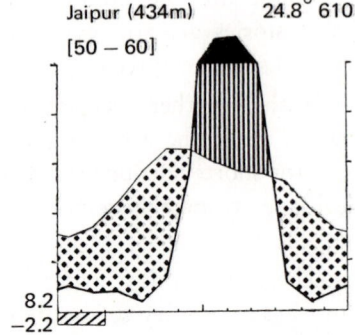

Jaipur (434m)
[50 – 60]
24.8° 610
8.2
–2.2

Fig. 25. Climatic diagram of Indian stations in the evergreen, semi-evergreen-, moist- and dry-monsoon-forest.

savanna) or short grass ("dry savanna") have replaced the original forest.

If the total annual rainfall sinks below 500 mm, the soil type becomes increasingly important for the natural vegetation. On deep, loamy sands the zonal vegetation is a climatically conditioned savanna, such as can be seen in Southwest Africa where the plant cover is still, to a large extent, natural. If the soil is stony, however, thorn bush predominates. In the region bordering on the desert the savanna

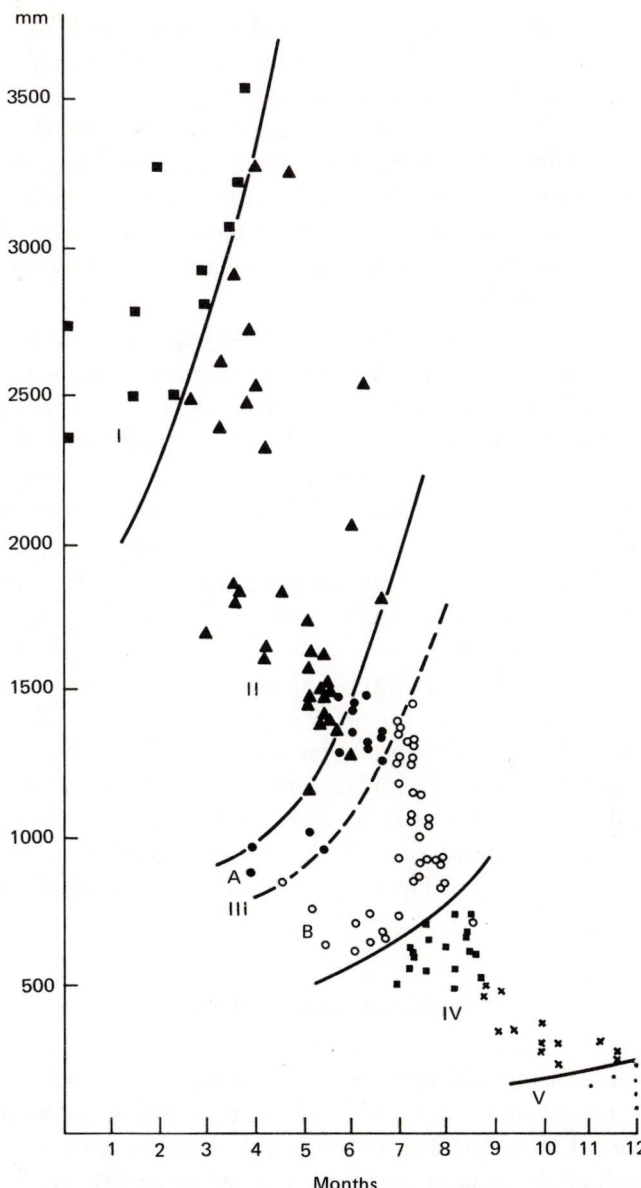

Fig. 26. The relation between forest vegetation and annual rainfall (ordinate) and length of drought in months (abscissa) in India. I evergreen- and II semi-ever-green-tropical rainforest, III monsoon forest (A moist, B dry), IV Savanna (thorn bush forest), V desert. Details in text (from a study carried out for UNESCO by H. Walter).

finally gives way to pure grassland and the thorn bush to a shrub semidesert. Such dry regions are unsuitable for cultivation but can support cattle and were formerly inhabited by enormous herds of big game. Over wide areas of Africa severe degradation has set in as a result of grazing and the annual burning of the grass. The population has steadily increased and, along with it, the number of cattle, so that it often happens that the natural plant cover cannot meet the demands of overstocking. Famine and distress for the population result from this misutilization.

What has so far been said does not hold true for Australia. Palaeotropical Indomalayan elements, which migrated to Australia via New Guinea only in recent geological time, form in the northeast the basis of a tropical evergreen rain forest that seems foreign to the typical Australian vegetation. Semi-evergreen deciduous forests are represented merely by small woodlands, also of Indomalayan origin.

In the rest of Australia autochthonous elements prevail, with the evergreen genus *Eucalyptus* as the predominant woody form. Although there are a few deciduous species of *Eucalyptus* (mainly *E. alba*) occuring on the fringes of the evergreen tropical rainforest, they occupy only an insignificant area. Apart from this, all of the moist and dry forest types in the summer rain regions, as well as the savannas, consist of evergreen *Eucalyptus* species. In the driest parts of central Australia these are replaced by evergreen species of *Acacia* with phyllodes. It is clear that, for these reasons, Australia's vegetation cannot readily be compared with that of other continents, although the physiognomic map in Fig. 13 gives no hint of its being of a special type.

2 Deciduous Tropical Forests

This type of forest is well represented in India and southeast Asia, although turned into cultivated land in many places. In Africa the areas denoted with 2 in Fig. 13 are largely anthropogenic savannas. Only south of the equator is the dry type of forest (Miombo) found stretching over enormous areas from East Africa as far as northern Southwest Africa (Fig. 27). In South America it is difficult to separate the deciduous forest from the "campos cerrados" with evergreen woody plants (p. 76).

Although the deciduous tropical forests occupy such a large area in the tropical zone, they have not been the subject of any detailed ecological investigations. The fact that the trees shed their leaves during the dry season is no new characteristic, since this phenomenon was already encountered in the evergreen forests with a seasonal rhythm. It is only the duration of the leafless state of the trees which has increased with the longer dry season: This duration depends upon

Fig. 27. *Colophospermum mopane* forest beginning to turn green at the commencement of the rainy season near the Victoria Falls in Rhodesia.

the water conditions in the habitat. In damp valleys trees can be found in full leaf while higher up on a dry slope they have already lost their leaves. The moist deciduous forests with their better reserves of soil water retain their leaves longer and lose them only toward the end of the dry season. In the botanical garden in Caracas a typical tree of the deciduous forest, *Hura crepitans* (Euphorbiaceae), remains green, flowering and fruiting during the whole of the dry season on regularly watered lawns, whereas on dry slopes it is bare for 3 to 4 months. This leads to the conclusion that leaf fall is facultative. Before the leaves are shed they appear to turn yellow rather than to dry out, which indicates that they probably behave like stenohydric xerophytes (see p. 90), although this would have to be proved by measuring the cell-sap concentration before leaf fall. Two ways are open to trees for surviving a dry period:

1. They may develop small, hard xeromorphic leaves that survive the dry season, just as do the sclerophyllous species of the winter-rain regions.

2. They may form large, hygromorphic leaves that are shed during the dry season and develop again in the rainy season.

Environmental conditions determine which of these two alternatives is the more economical. It is the total leaf area that is of primary importance to a tree. For the production of xeromorphic leaves about 2 to 3 times more organic mass per unit area is required than for the production of hygromorphic leaves. On the other hand, the evergreen leaves survive at least two vegetative periods and, water supplies permitting, they can also exploit part of the dry season for photosynthesis. Although hygromorphic leaves, which require far less material for their formation, are only productive for a short time, their photosynthesis relative to leaf surface is more intense. In summer-rain areas with a long dry season they generally appear to be the more economical as far as production is concerned (but not in *Eucalyptus*) and are successful in competition (12).

For the tropical deciduous forests the dry season is not a time of complete passivity since many tree species begin to flower toward the end of this season. Water losses by flowering, leafless trees are extremely low because the petals almost entirely lack stomata and transpiration is solely cuticular. It is probably a rise in temperature which provides the necessary impulse for the buds to open, since the temperature maximum is reached before the rainy season begins on account of the large amount of sunshine. This also holds true in a climate with diurnal fluctuations, such as in Venezuela, where this temperature increase amounts only to a few degrees; but it is known from the warm tropics with their uniform temperatures that plants as well as man react to very small temperature differences (p. 46).

As yet, organized forestry is seldom practiced in these moist deciduous forests, although much of their valuable wood is utilized, so that we still do not know much about their timber increment and primary production.

Frequent forest fires in the dry type of forest have led to a selection of fire-resistant species. The presence of grass undergrowth in the less densely wooded areas (savanna woodland) accounts for the readiness with which such areas catch fire.

Medina has determined the quantity of litter and the soil respiration over the course of an entire year in a small Venezuelan stand of deciduous trees 100 m above sea level (mean annual temperature = 27,1° C, rainfall = 1,334 mm). Soil respiration was three times more intense during the rainy season (May—October) than in the dry season. Litter production (without wood) amounted to 8.3 t/ha, and the mineralized dry matter, judged by CO_2 output of the soil, averaged 11.2 t/ha (11).

3 Climatic Savannas. Grasses and Woody Species as Antagonists

For the natives of South America the word "savanna" suggests a grassland area with or without scattered woody plants. For the purposes of geobotany, however, the word must be more strictly defined:

Only those ecologically *homogeneous grasslands upon which woody plants are more or less evenly distributed are termed savannas* (Fig. 28); if woody plants are altogether wanting, then these regions, in the tropics, are termed *grassland* (not steppes). Where a mosaic of grassland and smaller or larger stands of trees exists this is called

Fig. 28. *Acacia mellifera* ssp. *detinens* savanna in S. W. Africa, grass layer already dry following the rainy season.

parkland and consists of several ecologically distinct plant communities. In the past, the terms savanna and steppe have often been used very loosely, and difficulties have arisen from the general failure to distinguish between natural and anthropogenic plant cover.

The transition from open forest with grassy undergrowth (savanna-woodland) to true savanna is gradual. In the former, the tree-layer is predominant, whereas in the latter the grass is dominant (Fig. 68).

Grass and woody species are antagonistic plant types, the one usually excluding the other. Only in the tropics, where both summer rain and a deep, loamy sand coincide, are they to be found existing in a state of ecological equilibrium. The cause of the antagonism is to be sought in the differences in (1) their root systems and (2) their water economy.

1. Grasses possess a very finely branched, *intensive root system* with which a small volume of soil is densely permeated. Such a root system is specially suited to fine, sandy soils with an adequate water capacity in regions of summer rain where the ground contains plenty of water during the growing season.

Woody species, on the other hand, have an *extensive root system* with coarse roots which extend far into the soil in every direction, thus penetrating a much larger soil volume but less densely. This type of root system is well suited to stony soil where the water is not uniformly distributed, not only in summer-rain regions, but also in the winter-rain regions where the water which has seeped down to great depths has to be drawn up again by the roots in summer. For this reason grasses are unsuited to the latter type of climate.

2. As far as the water economy is concerned, it is characteristic of grasses that, given sufficient water, they transpire very stongly and photosynthesis proceeds very intensively. Within a short period of time their production is very large. At the end of the rainy season when water becomes scarce transpiration is not slowed down but continues until the leaves, and usually the entire aerial shoot system, dies. Only the root system and the terminal growing point of the shoot survive. Their meristem tissue is protected by many layers of dried-out leaf sheaths and is capable of surviving long periods of drought, even though the soil itself is practically desiccated. Growth recommences after the first rain.

The water economy of woody plants, on the other hand, with their large system of branches and many leaves is very intricate. At the first indication of water scarcity the stomata close and transpiration is radically reduced. If the lack of water becomes acute, then the leaves are shed, and during the dry season only the branches and the buds remain. Although well protected against loss of water, these have been shown to exhibit a very small but demonstrable loss of water over the course of several hours. The water reserves in the wood are insufficient to compensate for water losses over a lengthy dry period, which means that woody plants are obliged, under such conditions, to take up a certain, albeit it small, quantity of water. If the soil, however, contains no available water, then they dry out and die.

With these differences in mind, we are in a position to understand the ecological equilibrium of the savannas. As an example we have chosen an area in Southwest Africa with gradually increasing summer rainfall, uniform relief, and a fine, sandy soil which takes up all of the rainwater and stores the larger part of it (Fig. 29). We shall commence with the arid region where the annual rainfall amounts to only 100 mm (a) and the water cannot penetrate very far down into the ground. The roots of the small tufted grasses which permeate the upper, wet soil layers use up all of the stored water and then dry up at the end of the rainy season, with the exception of the root system and the apical meristem. *Woody plants cannot survive because the soil offers no available water during the dry season.* At a rainfall of 200 mm (b) the situation is similar: the soil is wet to a greater depth and the grasses are larger but still use up all of the water. Only when the rainfall reaches 300 mm (c) is some water left in the soil by the grasses at the end of the wet season, and although this is insufficient to keep them green, it is enough to enable small woody plants to survive the dry season and to form a shrub-savanna (Fig. 28). If the annual rainfall is 400 mm (d), then the larger amounts of water remaining in the soil at the end of the rainy season support solitary trees, and a tree savanna is formed. But it is still the grasses which are the dominating partner, and they determine how much water is left over for the woody plants.

Only when the rainfall reaches a level where the crowns of the trees link up to form a canopy whose shade prevents the proper development of grasses is the competitive relationship reversed. It is now the woody plants which are the dominant competitors in the

savanna-woodland or dry tropical deciduous woodlands, and the grasses are obliged to adapt themselves to the light conditions prevailing on the ground.

Such a labile equilibrium in the savanna is readily disrupted by man when he begins to utilize the land for grazing purposes. Water losses due to transpiration cease when the grass is eaten off, so that more water remains in the soil to the advantage of the woody plants (mostly *Acacia* species), which can consequently develop luxuriantly and produce many fruits and seeds. Their seeds are distributed in the dung of the grazing animals, and the tree seedlings are not exposed to competition from grass roots. The predominantly thorny shrubs grow so densely that thorny shrub-land is formed, which is then useless for grazing purposes.

In Texas mesquite *(Prosopis juliflora)* thickets have resulted from such grazing use.

In all grazing areas that are not rationally utilized there is a great danger of such a *brush encroachment,* and it is for this reason that thorn bush as a substitute plant community is nowadays more widespread than the climatic savanna. If the area is more densely populated, and if the woody plants are used for fuel or for making protective thorny hedges around the Kraals, then a man-made desert is often produced, which has only a covering of annual grass during the rainy period. The cattle starve during the dry season, with nothing but the straw remnants as fodder. Such is the situation, for example, in the Sudan. The only remaining natural savannas seen by the author were, besides those of Southwest Africa, in central Argentina at a rainfall of 400—200 mm with *Prosopis* as the woody species (see p. 177).

4 The Llanos on the Orinoco and the Campos Cerrados

Climatically conditioned, natural savanna cannot develop on a stony soil since this, as already mentioned, is unsuitable for grasses. Woody plants always dominate on this type of soil, becoming smaller as the rainfall decreases. On two-storied soils a special situation develops, as has been observed in Southwest Africa, where a bush-savanna is found with an annual rainfall of 185 mm (Fig. 30).

If the soil were deep and sandy, pure grassland would be expected at such a rainfall. In fact, the soil profile reveals that beneath the

10—20 cm of sand there is sandstone of the "Fischfluss" formation, which is arranged either in thin layers with small cracks or forms thick banks with larger crevices. The upper sandy layer cannot retain all the rainwater, and part of it seeps into the cracks in the sandstone. While the grasses can utilize the water from the sandy layer, the roots of the shrubs penetrate into the sandstone layer and take up the water from its cracks. The water reserves in the cracks of the

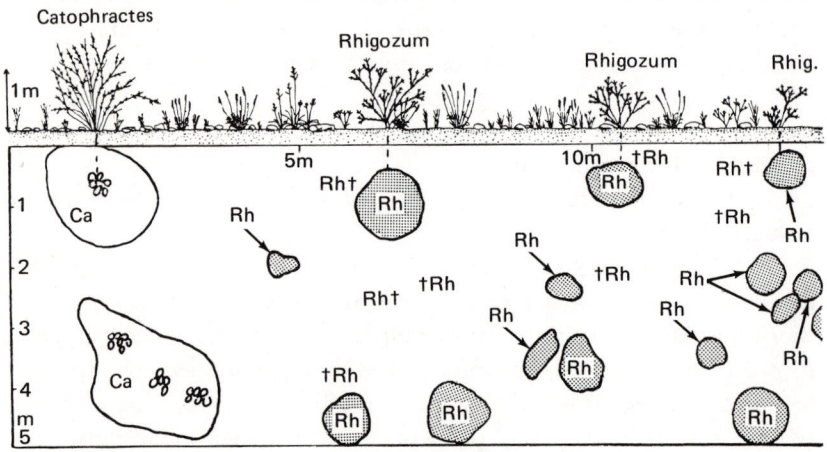

Fig. 30. Transect (1 m wide) through a typical patch of vegetation near Voigts-grund (S. W. Africa). Grasses dried out during the dry season. Below: ground plan of plant cover (without grasses): Ca *Catophractes*, Rh *Rhigozum* († dead).

thin-layered sandstone are sufficient only for the small *Rhigozum* bush, whereas the larger *Catophractes* shrub flourishes in the crevices of the thicker banked sandstone. The distribution of the bushes reflects the structure of the sandstone even in places where the covering layer of sand is missing. The bushes compete with one another, and although both types can germinate in the larger crevices, the smaller bushes are in time ousted by the larger ones and only their dead remains are left. There is however, no competition between grasses and woody plants.

This example ought to provide us with the key to the solution of the problem of the savanna in the llanos of the Orinoco. In the Tertiary period the llanos were a sea bed which in the Pleistocene filled up with alluvial deposits brought by the rivers flowing down from the surrounding mountains. This then became the gigantic plain of up to 400 km in width on the left bank of the Orinoco and stretching over 1 000 km through Venezuela, far into Colombia. It slopes almost imperceptibly toward the southeast. A distinction is made between the upper llanos (about 100 m above sea level) and the lower llanos, a few meters lower, which are flooded by the large rivers in the rainy season. We shall restrict ourselves to the upper llanos in the neighborhood of Calabazo. They are not flooded and have a reddish soil.

Fig. 31. Savanna with *Curatella americana* in the llanos near the Estacion Biologica Calabozo (Venezuela) during the dry season.

The climate has a typical dry season (see p. 33) with an annual rainfall of 1 300 mm, so that deciduous forests could be expected. These do indeed occur, with a typical floristic structure, scattered as small woodlands or even groves known as "Matas." Apart from this, the area is covered with grass about 50 cm in height, scattered with small trees *(Curatella, Byrsonima, Bowdichia)* and is in fact a true savanna (Fig. 31). Since the savanna cannot be climatic in origin (the rainfall is too high), edaphic factors such as soil conditions must be responsible.

The assumption has often been made that this is an anthropogenic savanna resulting from use of fire, but this explanation is too simple and uncritical since the savanna existed long before the arrival of the white man. It was neither cultivated nor used as grazing land by the

Indians. Fires caused by lightning are common occurrences in grassland, and although the Indians probably often burned down the dry grass, this was only possible because natural grassland already existed. Fire has certainly played its part in forming the savanna insofar as only fire-resistant woody species could survive in the grassland and on the fringe of the Matas, but it is by no means the primary reason for the existence of these enormous grassy areas. Edaphic conditions responsible for the absence of forests are not always identical. For

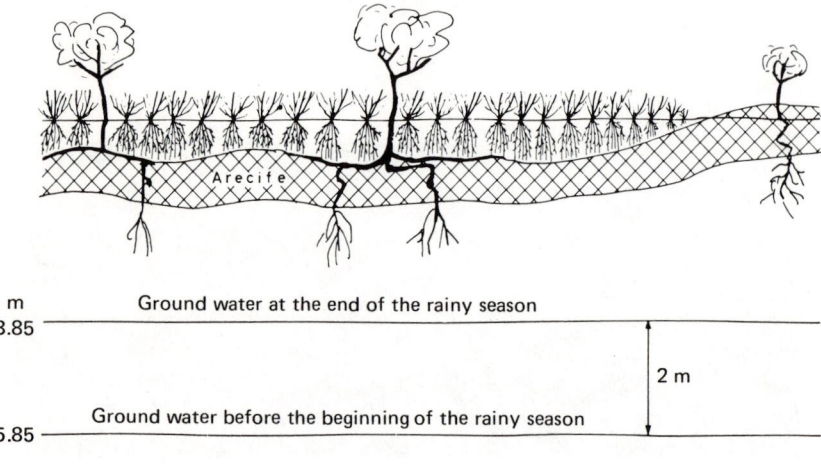

Fig. 32. Scheme to illustrate the situation in the llanos north of the Orinoco. Further details in the text.

the moment we shall limit our considerations to the central part of the llanos. It has been shown that at a time when the ground-water table was very high in this region a hard lateritic crust was formed (similar to slate) which was cemented by ferric hydroxide. This is called *"Arecife"* (Fig. 32).

Arecife runs beneath the surface at varying depths (mainly between 30 and 80 cm), rarely sinks below 150 cm, and can sometimes also be seen on the surface. The statement that Arecife is impermeable to water cannot be correct since the 750 mm of rain falling during only three months of the seven month summer rainy season cannot possibly be absorbed by the soil overlying the Arecife. Flooding would have to result in such a completely flat region, but this does not occur. The reddish color of the soil is another factor which speaks against prolonged water-logging. The ground-water table beneath the Arecife has in fact been found to rise from -575 cm to -385 cm, a rise of almost 2 m, by the end of the rainy season. Assuming a pore volume of about 50 percent for the alluvial deposits, this would mean that about 300 mm are retained by the soil above the Arecife and 1 000 mm seep through. The Arecife that has been laid bare by erosion on the river banks shows quite clearly that there are irregular channels penetrating the hard crust in places. This suggests that, just as in the example quoted on p. 72 for Southwest Africa, the grasses can take root in the fine soil above the Arecife and use up the 300 mm of stored water in the course of their development. Woody plants, however, grow wherever their roots can find a way through the Arecife layer to the damp, underlying layers where adequate water supplies are available. Groups of trees can grow wherever the channels through the Arecife are large enough or where a number of them occur close together. But small woodlands exist only in isolated patches where the Arecife is either entirely absent or extremely deep down. In such places deciduous forests (matas) are found, which are, in fact, the type of vegetation corresponding to the regionl climate. These savannas can thus be considered as a stable, natural plant community in which the distribution of the trees is a reflection of the Arecife structure. The following facts add weight to our point of view:

1. Wherever the Arecife is on the surface, a grass cover is completely wanting, but solitary trees grow on it at rather large intervals. The roots must in such cases reach down below the Arecife through existing channels.

2. *Curatella* remains green throughout the dry season and even produces new leaves and flowers, indicating that it is receiving a good supply of water. This holds for other woody species too. A small tree was found to transpire about 10 liters daily, and since the top soil is dry at this time, the water must have come from the layer below the Arecife.

Final proof would be afforded only by excavating the roots over larger areas, which would, however, be difficult to do. The use of dynamite on the Arecife would certainly favor woodland expansion. Further to the east, in the Mesa district, the Arecife is not so hard, but of a more compact, sandy nature, although still only penetrable in certain places by the roots of the trees.

In the extreme east there are white sandy patches, probably weathering products of the sandstone of the Guiana table mountains, which were deposited when the Orinoco still flowed through the Unare lowlands and emptied in the north. At the time of their deposition these white sands were so poor in nutrient material that forests were probably unable to develop upon them. For their phytomass they would have required more nutrients than were available in the soil. Grasses need less. A nutrient-poor soil is also the main reason for the widespread *"campos cerrados"* of the Brazilian shield, of which the Guiana table mountains are also a part, separated from it only by the low Amazon basin. Fertilization with trace elements has greatly increased the yield of crop plants wherever they are cultivated on the *campos cerrados*. Lack of phosphorus might also be responsible for this type of vegetation, just as in Australia. At least it would seem that it is not scarcity of water in this case which is of primary importance.

Savannas similar to those in the llanos are found on the slopes of the Cordillera del Interior in Venezuela. The soil here, too, consists of two horizons: a stony pavement with about 25 cm of fine soil underneath in which grasses can take root, and, further down, rocks with cracks in which *Curatella* and other small trees can root. If the rock is extensively permeated by cracks, then the trees grow close together. This is known as chaparral (*Curatella* = chaparra). But if the rock is more compact, then only grass grows on the slopes. The vegetation is partially the result of burning deciduous forests which formerly grew on deeper soil above the rocks. But after burning and subsequent grazing, severe soil erosion set in, the results of which are irreversible. Even today such an anthropogenic degradation of the vegetation can be observed.

The amount of water available to the woody plants during the dry season is always decisive for the different types of vegetation described, and this depends upon:

1. The amount of water stored in the soil at the end of the rainy period.

2. The rainfall during the dry period.

3. The duration of the dry period.

If conditions are favorable, forest develops on the slopes. If they are less favorable, chaparral is formed, or even pure grassland. This indicates that, apart from climate, soil conditions can play a decisive

role. The belief that the llanos savanna is original savanna is supported by its marked floristic similarity to those of Guinea and to `the
campos cerrados, which are definitely long-established plant communities. The margins of the Matas have in all probability been
pushed back by fire, and if they are completely protected from this
danger, the forest plants will advance somewhat. This does not, however, prove that the forest could completely usurp the savanna. Although statistics are not available, the possible role of fire caused by
lightning cannot be excluded (p. 171). Nowadays the grass is burned
before the rainy season with its thunderstorms commences.

5 Tropical Grassland on Alternately Wet and Dry Soils and on Flood Plains

Scattered over the savannas of the llanos in Venezuela are slight
depressions into which the water drains after a heavy downpour
(1961, 38 mm in 20 min.), and in which sediments of gray clay collect.
The water in the depressions reaches a depth of about 30 cm during
the rainy season, but toward the end of the dry season the gray
bottom dries out completely.

This alternating wetness and dryness is well tolerated by certain
grasses *(Leersia, Oryza, Paspalum, etc.),* but not by tree species other
than palm trees. This results in the formation of "Palmares," which
are grasslands containing the palm *Copernicia tectorum.* Such areas,
too, often burn, but the palms are resistant to fire (as are the tree
ferns) because they possess no damageable cambium (13). The dead
leaves sheathing the trunk are burned, and the outermost vascular
bundles are charred, but the resulting layer of carbon acts as an insulation against later fires. The apical meristem, surrounded by young
leaves, survives. If old leaves are completely missing from the trunk,
this is a sign that the palm-savanna has recently been burned. If they
sheath the trunk down to the ground, then the palm has not been exposed to fire at all. If only the lower portion of the trunk is bare,
this means that the palms has had a chance to grow for several years
since the last fire.

On permanently wet soils in Venezuela the palm *Mauritia minor*
is found. Black, acid peaty soils are formed, on which a few grasses,
or *Rhynchospora, Jussiaea (= Ludwigia), Eriocaulon,* and the insectivorous *Drosera* spp. (sundews) are found.

The llanos nearer the large rivers are 2 m lower than the others and are flooded for many months of the year. Trees are found only along the river levees as gallery forests.

In tropical Africa, too, grasslands are also widespread wherever the land is very flat. Disregarding the anthropogenic grassland, the following natural edaphic types of grassland can be distinguished in East Africa:

1. The grassland in the "Dambo," the flat watershed region between the Indian and Atlantic Oceans. From the air, the origins of the rivers can be detected in the wide, flat interconnected grassy strips on which rainwater collects in the rainy season and then flows away very slowly. These are typical temporarily wet habitats. Not until there is a noticeable slope is there a recognizable river bed cutting into the grassy areas. The latter gradually disappear, and at the same time the trees advance as far as the river banks.

2. The grassland in the depressions called "Mbuga," with black, sodium-rich soil which swells and is very soft during the wet season but dries out and is deeply cracked in the dry season. The soil contains little humus, but the color is due to a black humus-clay complex. Thorny *Balanites aegyptiaca* and the "flute" *Acacia, A. drepanolobium* flourish on such soils. The latter has large galls on the stipule-thorns, and these are riddled with holes bored by ants, so that as the wind blows past the openings it produces a peculiar sound.

3. Grassland that is inundated either by overflowing rivers or lakes. When such land is colonized by termites, patches of trees grow on the large, abandoned dome-shaped mounds and a *"termite" savanna* is formed, a mosaic of two different communities.

6 Tropical Swamps

High rainfall combined with relatively less potential evaporation accounts for the large surplus of water in the wet tropics. San Carlos de Rio Negro in Southern Venezuela, for example, with a rainfall of 3 521 mm has a potential evaporation of only 1 520 mm. In flat, poorly drained areas, therefore, extensive swamps have developed which, in Uganda, cover 12,800 km², about 6 percent of the entire country. The drainage basins of the river systems are not separated by watersheds but are connected with one another by a network of swamps. The flight from Livingstone to Nairobi provides excellent views of the large Lukango swamps as well as those surrounding Lakes Kampolombo and Bangweulu. The largest swamp area of all is formed by the White Nile in the Southern Sudan, which, together with its left tributary the Bar-el-Ghasal, fills an enormous basin lying 400 m above sea level. This region is known as the *"Sudd"* and extends 600 km from north to south and from east to west at its widest and longest. The total area covered is estimated at 150,000 km², and varies according to the water level. Half of the water of the Nile is lost by evaporation in the Sudd region. Seen from the air, the water appears to be dotted with small islands barely extending above its surface, but these are in fact swimming islands or floating lawns formed by the shoots of *Vossia* grass and *Cyperus Papyrus,* as well as grassy rafts of the South American *Eichhornia* and *Pistia.* From the air it is possible to distinguish free waterways and small stretches of water. When the water level sinks, part of the land emerges and grassland consisting of tall *Hyparrhenia rufa* and *Setaria incrassata* develops. The wettest parts are covered with *Echinochloa* species, *Vertiveria* and *Phragmites* (reed).

It was previously assumed that the Great Pantanal in the Mato Grosso in Brazil, bordering Bolivia and Paraguay, was a similar type of swamp, from which the southern tributaries of the Amazon and the right tributaries of the upper Parana arose. But this region is flooded only during the rainy season and is used as grazing land in the dry season, although ring-like lakes bordered with woodland remain. Swamp areas and watery basins are also widespread in the rest of the wet tropics. The aquatic vegetation consists of some cosmopolitan and pantropical species, but each region also has its own floristic peculiarities.

7 Mangroves and Shore Formations

Approaching from the sea a tropical coast with its protective coral reefs, one's attention is caught by partially submerged forests. At high tide the crowns of the trees barely extend above the surface of the salt water, and only at low tide are the lower portions of the trunks and the pneumatophores visible. Mangroves are found in the tidal zone in salt water, where the salt concentration is about 35 parts per thousand, 3.5% or 35%, which corresponds to a potential osmotic pressure of 25 atm. They are composed of about 20 woody species in all. Distinction must be made between the species-rich eastern mangroves on the coasts of the Indian Ocean and on the western coast of the Pacific, and the western mangroves on the east coasts of the Pacific and Atlantic Oceans, which are poorer as regards number of species. The best developed Mangroves are found near the equator in

Fig. 33. Outer Mangrove zone with *Rhizophora mangle* near Marina/Bahia de Buche (Venezuela).

Indonesia, New Guinea, and on the Philippines, but with increasing latitude they gradually become poorer until only one species of *Avicennia* remains. The last outposts are to be found at 30° N and 33° S (East Africa), 37—38° S (Australia and New Zealand), 29° S in Brazil, and 32° N on the Bermudas. Thus, although the Mangrove is at its best in equatorial regions, it also extends throughout the tropical and subtropical zones almost as far as the winter-rain regions or the warm temperate zone.

Mangroves are an azonal vegetation confined to the salt water of tidal regions. The chief genera of mangrove are *Rhizophora* with stilt roots and viviparous seedlings and the non-viviparous *Avicennia*, the white mangrove, which has thin pneumatophores growing up from horizontal roots spreading radially in the soil from the stem base. *Laguncularia* is a western mangrove. *Conocarpus* grows only where the salt concentration is low. In the eastern mangrove vegetation species of the genera *Bruguiera* and *Ceriops* (both viviparous with knee-like pneumatophores) occur, as well as *Sonneratia* (non-viviparous with thick pneumatophores) and species of *Xylocarpus*, *Aegiceras*, and *Lumnitzera*. The different mangrove species usually grow in distinct zones and only seldom are they mixed. This zonation is dependent upon the tides, since the nearer a species grows to the outer edge of the Mangrove, the longer and deeper does it stand in salt water.

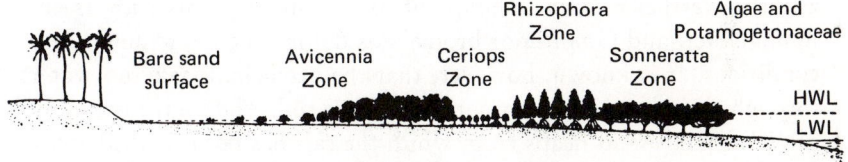

Fig. 34. Zonation of the east African coastal Mangroves (from Walter and Steiner). H.W.L. = High water limit, L.W.L. = low water limit.

The tidal range (the difference in depth between high and low tides) differs not only from place to place on coasts but also varies periodically according to the phase of the moon and position of the sun. It is at its maximum at new and full moon (spring tide) and at a minimum midway between the two (neap tides). The spring tides are at their highest twice a year when day and night are equally long (equinoctial spring tides).

Besides the coastal Mangroves growing on flat shores, often in a belt many kilometers wide and with no influx of fresh water from the land, there are also the estuary Mangroves, extensive in the delta regions of rivers, as well as the less common reef Mangroves growing on dead coral reefs protruding above the surface of the sea. The relatively simple ecological conditions in the coastal Mangroves have been investigated in most detail in East Africa, with special regard to salt relationships (Fig. 34).

The East African coast in the neighborhood of Tanga has a relatively dry monsoon climate. The potential evaporation is equal to or larger than the annual rainfall. Apart from a short dry season there is also a pronounced period of drought which is responsible for the fact that, within the tidal region, the further inland, and consequently the shorter the period of time during which the ground is flooded, the higher is the salt concentration of the soil. The most extreme conditions in the Mangrove zone are to be encountered at their inland margin, which is reached only by the equinoctial tides. Here the salt water in the soil is strongly concentrated by evaporation during drought, whereas during the rainy season the soil may be completely leached. Since no plant can tolerate such drastic variations in salt concentration, these areas are devoid of vegetation and are to be found wherever the climate prevailing at the inner boundary of the mangroves is such that a period of drought occurs. In Northern Venezuela, nevertheless, small clumps of salt-sensitive plants such as columnar cacti and *Opuntia* or bromeliads do, in fact, grow under such conditions. It is known, however, that the bromeliads take up water through their leaves and do not root in the soil. The cacti invariably grow on small sand heaps from which the salt has been washed away in the rainy season. They obtain their water by means of shallow roots spreading in the sand and are therefore unaffected by the underlying salty soil. The tissues of both cacti and bromeliads are free of salt, and they are not halophytes — a further example of the observation that obvious soil properties may not always be an indication of the ecological conditions under which plants grow.

In very humid regions, on the other hand, the exposed areas are continually being washed by rainwater so that the salt concentration of the water in the soil must decrease landward. This is also true, moving upriver, of the estuary Mangroves which, via a brackish zone (where the fern *Acrosticum*, the *Nipa* palm, *Acanthus ilicifolius*, and many other species are found), give way to a freshwater community without the interpolation of a barren zone.

Plants rooting in saline soils take up a certain amount of salt and store it in the cell sap. In the case of mangroves the salt concentration of the cell sap in their succulent leaves is roughly equal to that of the soil. Apart from this, nonelectrolytes are present in a concentration usual for tropical species. Fig. 35 illustrates the typical zonation and the potential osmotic pressure in the soil and the mangrove leaves.

Zonation in the Mangroves results from competition between the various species. Investigations carried out in East Africa suggest that the salt factor is decisive in determining this zonation. *Avicennia* is the weakest competitor but has the highest salt resistance, so that stunted individuals of this species constitute the landward limit of the Mangrove. *Sonneratia* is the strongest competitor but seems not to tolerate an increase in salt concentration above that of seawater and is therefore confined to the outer fringes. In continuously humid areas the situation is more complicated: *Avicennia* appears to be confined to sandy ground, while *Sonneratia* prefers silty soil. In such regions the type of soil, duration of inundation, water movement, decrease in saltiness, or variations in salt concentration seem to be of greater significance.

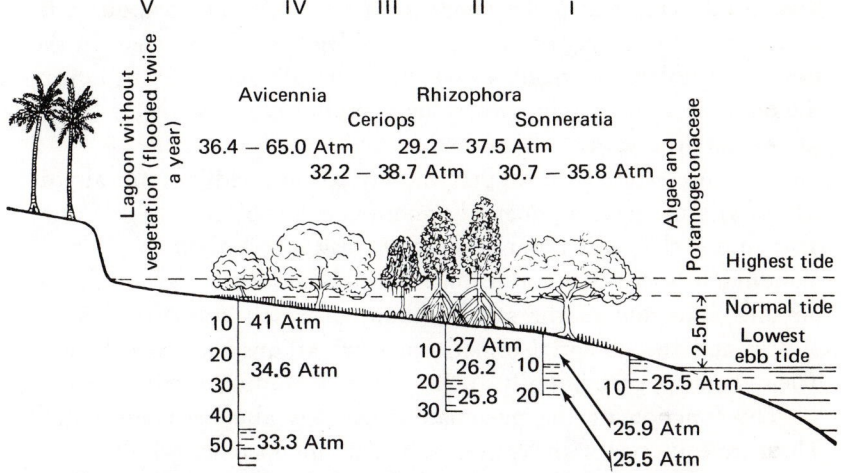

Fig. 35. Concentration of the cell-sap in atm (smallest and highest) of the leaves of mangrove species and of the soil solutions at various depths (in cm). Coastal Mangroves of E. Africa.

The salt economy of the mangroves presents an interesting problem. Mangroves are unable to take up seawater in its original form because the salts which would be left behind after the loss of water by transpiration would soon lead to the formation of a saturated salt solution in the leaves. It has recently been shown by direct measurement that suction forces of from 30 to 35 atm can be produced in mangrove leaves. This is higher than the osmotic pressure of the soil solutions. These suction forces are transmitted to the roots by the cohesion tension in the conducting vessels. The roots act at the same time as an ultrafilter, permitting only the passage of almost pure water which is then transported to the leaves. Only the very small amount of salt necessary for producing the suction tensions enters the plants, and this is then stored in solution in the vacuoles of the leaf cells.

The mechanisms whereby the salt concentration is regulated have not so far been completely elucidated. Some of the salt from the old, fading leaves can be transported to the young, developing leaves, or an excess of salt can be eliminated when the old leaves drop. *Avicennia* has salt glands, situated on the underside of its leaves, for regulating the salt concentration. This species can excrete a 4.1 percent salt solution, which is more concentrated than seawater; the proportion of NaCl to KCl is the same as in seawater (90 and 4 percent). No excretion takes place at night, but it reaches a maximum at midday. In 24 hours 0.2—0.35 mg of salt per 10 cm² leaf area is excreted. In the dry season the salt crystallizes on the underside of the leaves during the daytime, and at night, when the humidity is greater, the salt dissolves and drips away.

It is interesting to note that the viviparous seedlings are almost free of salt and have a potential osmotic pressure of only 13—18 atm. This means that water must somehow reach them with the help of glandular tissue in the cotyledons. But as soon as the seedlings drop off and take root in the salty ground, their salt concentration increases and the potential osmotic pressure attains the normal level. The radicle appears, initially at least, to be permeable to salt.

The function of the pneumatophores has also been elucidated. They are equipped with lenticels with minute openings which permit only the entry of air but not of water. When the pneumatophores are completely submerged in water, their intercellular oxygen is used up in respiration and a negative pressure develops because carbon dioxide, being very soluble, escapes into the water. As soon as the roots emerge from the water, a pressure compensation takes place, whereby air (and oxygen) is drawn in. The oxygen content of the intercellular spaces in such roots therefore varies periodically between 10 and 20 percent.

Together with their fauna — the numerous small fiddler crabs and the mangrove fish *Periophthalmus* which can be seen crawling out of water and up the trees — the Mangroves present a highly interesting community, belonging neither to the sea nor to the land.

The sandy shore formations of tropical coasts offer fewer peculiarities. Beyond the barren zone resulting from exposure to the force of the breakers, sand plants with long runners can be found, among them the widespread *Ipomoea pes-caprae* and the halophytic *Sesuvium portulacastrum* and *Sporobolus virginicus*.

Still further inland, beyond the influence of the salt water, the sand in the tropics very soon becomes covered by shrubs and trees, the floating fruits of which can be seen in the drift on all tropical shores. Typical representatives are *Terminalia catappa* and coconut palms, although nowadays the palms are nearly all planted by man. *Barringtonia, Calophyllum, Hibiscus tiliaceus,* and *Pandanus* are characteristic for the eastern oceans, and *Coccoloba uvifera* (Polygonaceae), *Chrysobalanus icaco,* and the poisonous *Hippomane mancinella* (Euphorbiaceae) for the western oceans. No large dune areas occur in the tropics, with the exception of the north coast of Venezuela, near Coro, where a semidesert type of climate prevails. As a result of the continual trade winds blowing in from the northeast or east-northeast, large quantities of sand drift landward from the beach and are trapped by *Prosopis juliflora*. The dunes thus formed continue to grow in the direction of the wind and are soon covered by *Prosopis* bushes. In this manner a series of dune ridges is established, running parallel to one another and to the wind direction and reaching a considerable height. Migrating dunes or barchanes are found in one part of the dune region, probably resulting from wood clearing. They join up to form ridges at right angles to the wind direction.

III Subtropical Semidesert and Desert Zones

1 The Water Supply of Plants in Arid Regions

Regions in which the potential evaporation is much higher than the annual rainfall are termed arid. These regions can be further subdivided into semiarid, arid, and extremely arid. The cold winter period which is typical of the arid regions of the temperate zone is lacking in the subtropical zone (see chapter VII). The extreme dryness of the arid regions has led to the assumption that desert plants must possess special physiological properties — a physiological resistance to drought — enabling them to grow under such conditions. In particular, the allegedly high cell-sap concentration is often mentioned in connection with the ability of the plants to take up water even from almost dried-out soils. However, detailed ecophysiological investigations over the past decade have shown that this view is incorrect. The water supply of desert plants is not so poor as would be suggested by the low rainfall. Rainfall measured in millimeters is equivalent to liters of water per square meter of ground surface. In order to judge how much water is available to the plants, the transpiring surface per square meter of ground surface must be calculated.

Although there are many different kinds of deserts, they are all alike in the sparseness of their plant cover, so that the character of the landscape is determined by the naked rock and not by the plants. In order to study the exact relationship between rainfall and density of vegetation, identical life forms must be compared (e. g., grasses or trees with a similar kind of foliage) and a region must be chosen in which the rainfall varies over a relatively short distance while the temperature conditions remain more or less constant. Furthermore, climatopes with similar soils should be chosen where the vegetation has in no way been disturbed by human action.

Suitable regions are to be found in Southwest Africa with grass cover and an annual rainfall of 100—500 mm, and in Southwest Australia with Eucalyptus forests and a rainfall of 500—1,500 mm. Such investigations have revealed a linear relationship between the amount of rainfall and the production of plant mass or transpiring area (Fig. 36). This also holds true for the creosote bush desert (*Larrea divaricata*) in Southeast California (15).

From this it can be deduced that *the water supply per unit of transpiring surface is more or less the same in arid and in humid regions* (annual rainfall of 500—1,500 mm). The drier the region, the further apart the plants grow, thus leaving a greater area from which the individuals can take up water. This has been confirmed in North Africa in olive plantations. The number of trees per hectare decreases proportionally to the decreasing rainfall until, finally, only 25 trees per hectare are left. But since the individuals bear approximately the same amount of fruit, it is evident that the water supply is unaltered. In cereal-growing areas, too, it is known that crop plant density decreases with decreasing rainfall.

To take up water from a larger soil volume, a correspondingly larger root system has to be developed. Combined with a steady reduction of the transpiring surface, this larger system ensures an adequate water supply in the face of increasing aridity. A negative water balance in the plants and a rise in cell-sap concentration have been shown to be accompanied by an immediate and marked inhibition in shoot growth, while root growth is at first even enhanced (2). Whereas in wet regions the larger part of the phytomass is aboveground, in arid regions it is underground. This does not mean that the roots

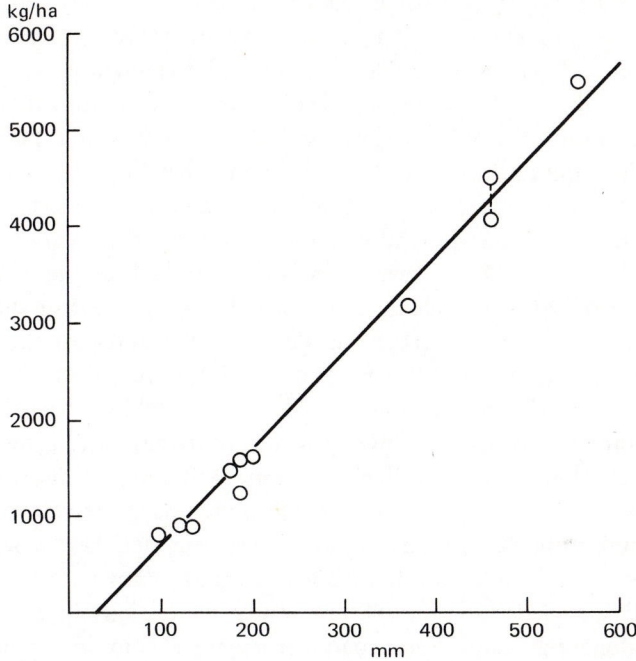

Fig. 36. Production (aboveground dry mass in kg/ha) of the grassland of S. W. Africa in relation to annual rainfall in mm.

penetrate deeper in dry regions, as is usually suggested. Rather the opposite is true: the root system flattens out. The scantier the rainfall, the less deeply can the soil be wetted, and beneath the upper, water-containing layer there is no water available at all for the plants. Long tap roots have been observed only in plants that are dependent upon ground water. These are, however, special cases from which generalizations should not be made.

In extremely arid regions with a rainfall below 100 mm a change in type of plant cover is noticeable. Instead of being *diffuse* or evenly distributed over apparently flat areas the vegetation changes to a so-called *contracted type,*[1] the plants growing only in the often barely noticeable erosion gulleys or depressions, while the raised areas have no vegetation at all. This is connected with the distribution of the water in the soil.

In extreme desert regions the soils, apart from the shifting sands, usually have a surface crust which can be moistened only with difficulty. The rare but usually torrential rain therefore scarcely penetrates the soil but to a large extent runs off the surface. The sandy erosion gulleys and depressions therefore receive much more water than the rainfall would suggest, and this run-off water penetrates deeply into the soil. In such places the plants develop roots that reach down as far as there is any water, and this is sometimes several meters. In some places ground water even collects in the valleys. With a rainfall of only 25 mm vegetation is present in all of the valleys in the desert near Cairo-Heluan. Assuming that 40 percent of the rain-water runs off into the deeper parts of the relief and that these depressions account for only 2 percent of the total area, then, at a rainfall of 25 mm, the same amount of water is available to the plants in these biotopes, thanks to inflowing water, as if they were growing on level ground with a rainfall of 500 mm. Water losses due to transpiration of the plant cover in such a habitat near Heluan were, in fact, found to be 400 mm. Even in a rainless summer the cell-sap concentration of these plants does not rise, thus confirming that they are well supplied with water. The sandy depressions in the gravel desert along the Cairo-Suez road permanently contain 2.5 percent water at a depth of only 75 cm (wilting point 0.8 percent), so that they never dry out and are capable of supporting a sparse perennial vegetation. In some erosion gulleys roots descend as far as 5 m, depending upon the depth to which the soil remains damp. Despite extreme aridity 200 vascular plant species can be found in the neighborhood of Cairo.

The water supply of plants in extreme deserts is therefore not so poor as is usually assumed. However dry the soil surface may appear to be, there is always some water available, at least at certain times, wherever plants are found growing in the desert. These plants must, of course, be able to resist long periods of drought, and this they

---

[1] Végétation contractée of Monod.

achieve chiefly by means of special morphological adaptations. There is no significant plasmatic resistance to drought, and the cell-sap concentration is generally low.

The berber population of southern Tunesia has exploited the principle of "contracted vegetation" for countless centuries in order to obtain crops at a rainfall of only 200 mm and less annually. Each small gulley is provided with a dam to prevent the water from draining off, and date palms or barley can be cultivated on the damp soil.

It has been established that a similar type of *run-off farming* was practiced in pre-Arabian times in the Negev desert. The old dams have been renovated in recent times, and experiments with various cultivated plants have been successful.

2 Ecological Adaptations of Plants in Arid Regions

All plants growing in arid regions have been called "xerophytes." This is incorrect, because in every such region there are habitats, such as the oases, where the plants are always well supplied with water. Species typical of the humid tropics even grow in such habitats. In the rainless Aswan desert, on an island in the Nile, coconut palms, mangoes, papaya, mate, sweet potatoes, manioc, camphor trees, mahogany trees, coffee, pomegranates, and many other species typical of the Indian monsoon forests are cultivated with the aid of artificial irrigation. The microclimate in the dense plantations is less extreme than in the open desert. In dry valleys with ground water plants can grow under natural conditions without suffering any water shortage and without any sign of an adaptation to the dryness. Besides most deserts have at least a brief wet season, with the exception of the rainless central Sahara, the Namib, and the Peruvian-Chilean desert. The species which develop during these damp periods *(ephemerals)* including those which survive the rest of the time as seeds (therophytes) or in the ground (geophytes) do not exhibit any particular adaptation to water shortage either.

Poikilohydric species are closely related to the ephemeral plant species. The former retain their vegetative organs during the drought, can stand complete desiccation without undergoing any damage, and after a shower they are again fresh and active. This group includes lichens, which are characteristic of fog deserts, some mosses belonging to the dry regions, and certain ferns with hard leaves or with scales. This group avoids, however, the extreme arid regions. Among the angiosperms the dwarf bush *Myrothamnus flabellifolia* (Rosales) deserves mention.

The remaining species capable of surviving drought without becoming desiccated can be termed *xerophytes,* of which the *succulents,* with special water-storage organs, form a group of their own.

The xerophytes can further be distinguished according to their ecological behavior into three subgroups connected by transitional forms:

1. *Malakophyllous xerophytes,* which are characteristic of semiarid regions. They have soft leaves which wilt under dry conditions while the cell-sap concentration rises steeply. They lose their leaves in lengthy dry periods, and only the youngest of the leaves within the hair-covered buds survive. Typical examples are the many Labiatae and Compositae of arid regions, *Cistus* spp. among others.

2. *Sclerophyllous xerophytes* with small, hard leaves, owing their rigidity to mechanical tissue. They are found especially in regions with a long summer drought and are able to reduce their transpiration to a minimum when water is scarce, whereby the cell-sap concentration rises only in extreme circumstances. Typical examples are provided by evergreen oaks, and olive trees, etc.

3. *Stenohydric xerophytes,* which are able to prevent a rise in cell-sap concentration by shutting their stomata at any sign of water shortage. Gaseous exchange and photosynthesis are also of necessity brought to a standstill so that the plants are in a state of starvation. The leaves of such species do not dry out during the long droughts but turn yellow and finally fall off. Some nonsucculent *Euphorbia* species may be cited as examples, although many plants of the extreme deserts also belong to this group. Since there is no competition among the aerial parts of the plants in the deserts, the only important thing is for them to survive the drought and not to produce large quantities of phytomass. This they achieve with incredible endurance, often as pitiful-looking cripples. Sometimes they even manage to achieve a great age, as for example the small *Eurotia ceratoides* in the Pamir, which can reach the age of 300 years. Xerophytes require the presence of some available water in the soil during the dry period, and in this they differ from succulents.

Succulents are water-storing species which use their stored water extremely sparingly in times of drought. Their small absorbing roots die, so that no water is taken up at all from the ground during the dry period. According to the nature of the organ responsible for storing water during the rainy season, the succulents can be divided into:

1. Plants with succulent leaves, such as *Agave* and *Aloe,* or *Cotyledon, Crassula,* and *Sanseviera.*

2. Plants with succulent stems, such as cacti and many species of *Euphorbia* and *Stapelia.*

3. Plants with succulent roots, that is to say, with underground storage organs, such as *Asparagus* species, *Pachypodium succulentum,* as well as some Leguminosae with enormous tubers to be found growing in the sandy regions of the Kalahari.

The concentration of the cell sap in the succulents is very low and does not rise even during the long periods of drought when large amounts of water have been lost. The water content calculated on the basis of dry weight remains constant since respiration involves the breakdown of organic compounds. Many succulents can survive for a year without taking up any water, and in many of them the *de Saussure effect* has been demonstrated. This means that they open their stomata at night, when water losses due to transpiration are small, and take up CO_2, which leads to the formation of organic acids and thus to a rise in acidity of the cell sap. The stomata are closed in the daytime, and the CO_2 which has been bound during the night can be assimilated in daylight, with an accompanying decrease in acidity. In this manner the necessary gaseous exchange is effected with a minimum loss of water (see p. 51).

The salt plants or *halophytes,* constituting a very important group in many deserts, must also be mentioned. Their occurrence depends upon a saline soil rather than upon the climate, and for this reason the salt factor will now be discussed. Halophytes usually exhibit succulence, although they should not be classed with the true succulents. Their succulence results from an intense storage of salt, that is to say of chloride, with the result that the *cell-sap concentration is often very high* and may even exceed 50 atm. The Mesembryanthemums represent a link between the true succulents, with their low cell-sap concentration, and the halophytes. They can be extremely succulent and can also occur on nonsaline soils, although their cell sap always contains a certain amount of chloride.

3 Saline Soils in Arid Regions

As a result of the long periods of drought in arid regions, the rivers carry water only periodically or even sporadically. Since the potential evaporation is many times larger than the annual rainfall, the depressions are often devoid of any outflow, whereas in humid regions every depression is full to overflowing. The water running down into such depressions in the arid regions evaporates, and the salts which are left behind concentrate over a period of time, so that a saturated brine may even be formed, from which salt then crystallizes out. The largest part of the salt is NaCl, but Na_2SO_4, $MgCl_2$, $MgSO_4$, and other salts also occur. Calcium is rapidly precipitated as $CaCO_3$, but $CaSO_4$ or gypsum is somewhat more soluble. Gypsum can however, occur in crystalline form in soils. In the process of weathering of silicates and of clay formation, sodium ions are set free. Although almost 20 g of chloride ions are present per liter of seawater (sulphate ions only 2.7 g) chlorine-containing minerals are extremely rare, so that chloride ions are not set free by weathering. In river water, however, the presence of NaCl can always be demonstrated. The NaCl of the salty soils of arid regions can be of various origins:

1. Salt from rocks that were formed as marine sediments. This salt can be washed out of the rocks by rainwater and carried into the undrained depressions. In deserts with marine sedimentary rocks of Jurassic, Cretaceous, or Tertiary age, for example the northern Sahara and the Egyptian desert, saline soils are common, whereas in arid regions with underlying magmatic rock or terrestrial sandstone hardly any saline soils are found at all.

2. Arid regions that in the most recent geological past were lake or marine beds are also brackish. Examples are the areas surrounding the Great Salt Lake in Utah, around the Caspian and Aral seas in Middle Asia (see pp. 182 and 185), and the Tuz Gölü in central Anatolia.

3. Along the arid coasts with heavy surf the seawater is turned into a fine spray by the force of the breakers. The small droplats dry and form a salty dust which can be blown inland. This salt is then either washed into the soil by rain or fog, or it is simply deposited. A similar process also takes place in humid regions, but here the salt is continuously being washed out and returned to the sea via the rivers (cyclic salt), In arid regions with no outflow, however, the salt concentrates, and in this way leads to such brackishness as is encoun-

tered in the outer Namib desert and the arid parts of western Australia. Once salt accumulates on the surface of depressions, it can be blown still further by the wind.

4. Brackishness can also occur if salty water comes to the surface in springs, as it does in the northern Caspian Lowlands. In this case the salt originates in marine beds which desiccated in earlier geological times and formed large salt deposits at a considerable depth.

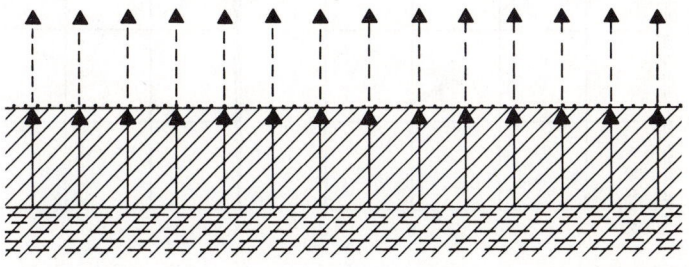

Fig. 37. Formation of a salt crust due to capillary ascent (continuous arrows) of the ground water (horizontal dashes) and evaporation of the water (broken arrows).

In fact, all of the chloride of the salt found in the deserts came from the sea, where it was deposited in the course of the Earth's history, the chlorine being mainly from volcanic exhalations containing HCl.

In the desert, salt is continuously being washed from the higher elevations to lower-lying parts, so that mostly only the depression soils are saline. But if the sedimentary rocks contain much salt and rainfall is very scanty, as it is around Cairo-Heluan, then the soil of the plateau habitats also contains salt. There is no salt transport in the completely rainless central Sahara, and the lower areas receive no additional salt.

As far as plants are concerned, it is the salt concentration of the solution surrounding the roots that is important and not the salt content of the soil calculated on a dry weight basis. In slightly salty but dry soils the concentration is often higher than in very saline but wet ground.

Evaporation from the soil surface in places where the ground water is less than one meter below the soil surface can also lead to salt

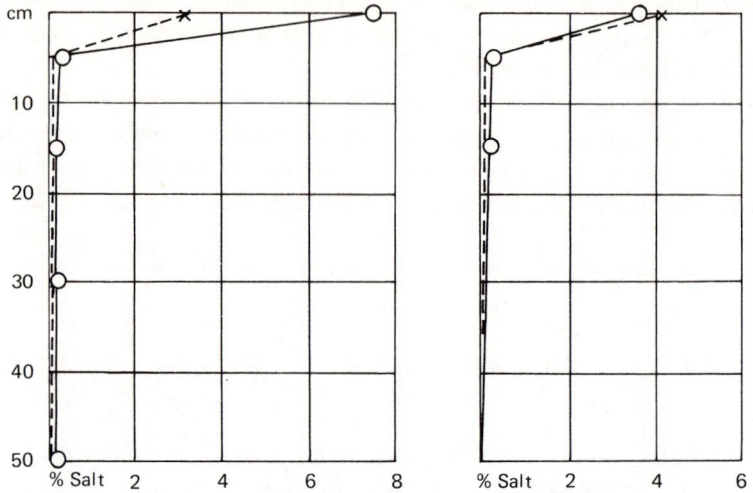

Fig. 38. Salt content at various depths in a watered plot (left) with ascending ground water and in an unwatered plot in the Swakop Valley of S. W. Africa. NaCl = full lines, Na_2SO_4 = broken lines. The salts collect only at the surface.

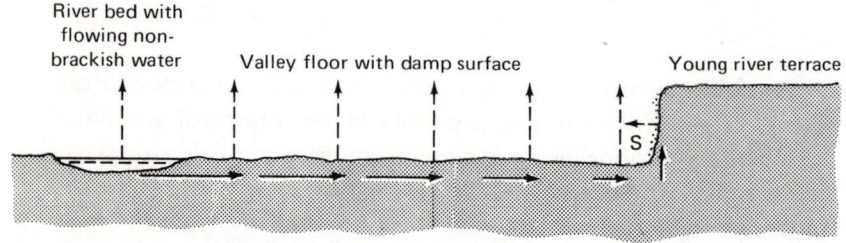

Fig. 39. Salt accumulation in Swakop Valley (Namib, S. W. Africa). The arrows indicate the direction and magnitude of water flow in the ground, the broken arrows the evaporation. The salt concentration increases towards the sides of the valley; salt efflorescence occurs at S at the foot of the terrace, where the water flow ends.

accumulation. Water rises to the surface by means of capillary forces and a salt crust forms (Fig. 37), even if the ground water contains only minute quantities of salt (Fig. 38). A salt crust forms wherever the capillary water column ends, which is at the highest point of the microrelief (Fig. 39). The occurrence of a salt crust in the dry periods does not interfere with plant growth as long as the roots have access to nonbrackish ground water.

If irrigation in arid regions is not accompanied by at least a certain degree of drainage, the cultivated areas necessarily turn brackish in time, even if the water used for irrigation purposes contains only a small quantity of salt. Extensive cultivated areas in Mesopotamia, in the Indus region, in California and other parts of the western U.S. have been transformed into salt deserts in this manner. So far, this has not happened in the undrained cotton fields of the Gezira in the Sudan, but only because the water of the Blue Nile, which is used for irrigation, contains hardly any salt at all. Small quantities of salt are removed in the crop itself with each harvest.

4 The Salt Economy of Halophytes

Plants growing on saline soils have been termed "halophytes." It would be preferable, however, to take the plant itself as a starting point in making any definition. *True halophytes are plants that store large quantities of salt in their organs without thereby undergoing any damage.* They may even benefit from the salt if its concentration is not too high. The salts involved are usually NaCl, but can occasionally also be Na_2SO_4 or organic Na salts.

What was said for the mangroves also holds true for all halophytes. The osmotic effect of the salt concentration in the soil has to be balanced by an equally high salt concentration in the cell sap. Since there are also other osmotic substances present in the cell sap, the transpiring organs can produce a suction tension sufficiently high to extract water from saline soils. The salts in the cells have an effect upon the protoplasm and are toxic to salt-sensitive species, which for this reason cannot survive on saline soils. Facultative halophytes that can tolerate salt develop, nevertheless, better on a non-saline soil. But the growth of true or euhalophytes is, in fact, stimulated to a certain degree by salt. On normal soils containing only traces of NaCl, such plants greedily take up the available salt so that their salt content remains high. The stimulating effect is due to the chloride ions which cause a swelling of the proteins and, therefore, lead to a special ionic hydration of the protoplasm: This results in a cell hypertrophy due to water uptake or, in other words, a succulence of the organs. Only the chloride ion has this effect, while the sulfate ion has the oppsite effect. Certain halophytes that store larger quantities of sulfates as well as chlorides in the cell sap are only barely succulent or are not succulent at all. A distinction must be made, therefore, between the *chloride- and the sulfate-halophytes,* although they can exist side by side on the same type of soil. The uptake of salt is thus seen to be species-specific. Investigations on halophytes are quite clearly incomplete if only the salt content of the soil is determined since, for the plants, only those salts are important that come into contact with the protoplasm. *The concentration and composition of the salts in the cell sap must therefore always be measured.*

Even the euhalophytes have an upper limit, varying from species to species, for the concentration of salt tolerated in the cell sap. If this rises too far the plants wilt, which in the case of the Chenopodiaceae is usually accompanied by their turning red due to the formation of N-containing anthocyanins, and finally die. In yet a further group of halophytes the sodium in the cell is at a higher equivalent concentration than that of the Cl and SO_4 put together, so that the Na ions must be equilibrated by anions of organic acids. When such plants die, carbonic acid is produced as a result of the breakdown of organic acids and the sodium reaches the soil in the form of Na_2CO_3 (soda), which leads to an increased alkalinity. These plants are termed *alkalihalophytes.*

There are also *salt-excreting halophytes,* usually nonsucculent species with salt glands, as, for example, *Limonium, Frankenia, Reaumuria,* and halophilic grasses. The well-known tamarisk tree *(Tamarix),* of which there are many species in arid regions, also has salt glands. Salt dust rains down if the branches of these trees are shaken.

Most of the halophytes in arid regions, growing on the wet saline soils of the salt pans *(hygrohalophytes),* are more concerned with their salt economy than with the water supply itself. But there are others that grow on saline soils and often suffer from water shortage, despite a considerable salt concentration in the cell sap *(xero-halophytes).* *Atriplex* species, *Haloxylon* and *Zygophyllum* are of this type, and they often reduce their transpiring surface during the dry season. *Zygophyllum,* for example, sheds its small leaves, while others lose the young terminal shoots or even the green bark of the leafless shoots from the preceding year.

Fig. 40 shows the large difference in composition of the cell sap of halophytes and nonhalophytes.

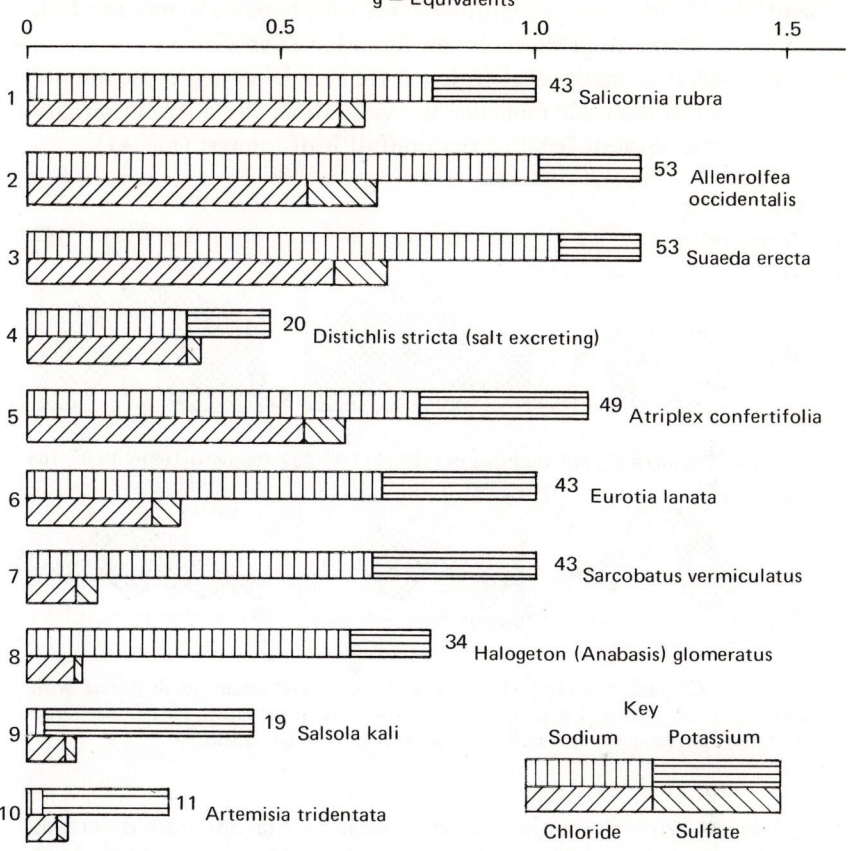

g — Equivalents

Fig. 40. Concentration of the inorganic salts in the cell sap of the transpiring organs of halophytes. Figures on the right are concentrations (in atm) of the salt solution in the cell sap. 1—5 chloride-halophytes (all succulent except the salt-excreting grass *Distichlis*); 6—9 alkali-halophytes (cations bound to organic acids, often oxalic acid); 9 leads on to the non-halophyte 10 (almost only potassium as cation). Samples taken in 1969 by the author in the region of the Great Salt Lake, the analyses carried out by Utah State University in Logan.

5 Different Types of Climate in the Arid Regions

The term desert is a relative one. Coming from the humid eastern part of North America the Southwest looks like desert, although Tucson (Arizona) has an annual rainfall of 300 mm. On the other hand, the Mediterranean coast of Egypt, with barely 150 mm rainfall, is not considered to be desert by an Egyptian from Cairo.

In general, a subtropical region is termed desert when the annual rainfall is less than 200 mm. For the vegetation, the distribution and not only the absolute level of the rainfall is of interest (Fig. 41).

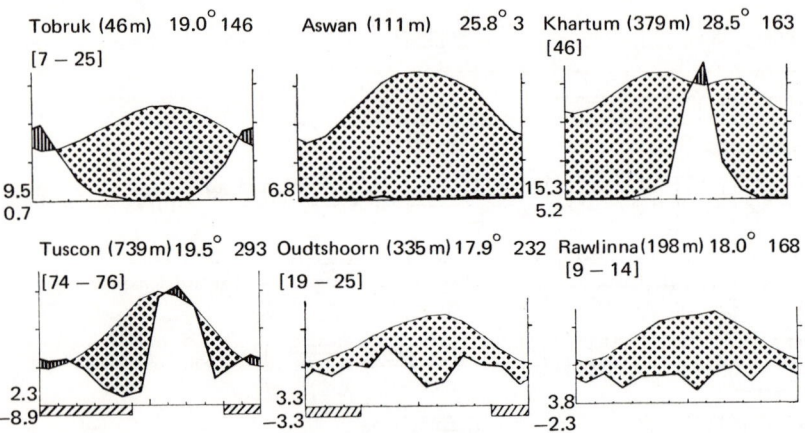

Fig. 41. Climatic diagrams of desert stations. Above, from North Africa with winter rain, no rain and summer rain; below, with two rainy seasons (Sonoran desert and Karroo) and some rain at all seasons (Rawlinna, Australia).

Deserts with two rainy seasons, such as the Sonoran desert of Southern Arizona and New Mexico and the Karoo and Namaland of South Africa, particularly favor the development of succulents. The northern Sahara and the Mohave desert of California have a winter rainy season, and only the central Sahara is rainless in a strict sense. Summer rains occur in the southern Sahara, in the Sudan, and in the Sind or Thar desert, east of the lower Indus. The inner Namib also has scanty summer rainfall.

The Chilean-Peruvian desert and the outer Namib of Southwest Africa border upon coasts with cold currents. Rainfall is extremely rare, but fog is common (on approximately 200 days of the year in the Namib). This fog influences the vegetation only in places where it is forced to rise up mountain slopes, thus producing large quantities of condensation water; but on the flatter coasts, such as Senegal and Mauretania, it does not provide a source of water.

The situation in Australia, however, is peculiar. The interior of this continent and, above all, its west coast are arid. From the south the influence of winter rains is noticeable while summer rains approach from the north. But the continent is too small for there to be a complete separation of the different rainy regions such as occurs in the Sahara. Winter- and summer-rains overlap, therefore, with the result that in the driest areas, with a rainfall of only 100—150 mm annually, the rain may fall at any time of year. This explains why the curve of long-term mean monthly rainfall in the climatic diagram is more or less flat. In the south the winter rains predominate, in the north the summer rains predominate.

6 Soil Texture and Water Supply of Plants

The quantity of rain is only of indirect importance to the plants of arid regions. The amount of water remaining in the soil, and thus available to them, is of far greater importance. Part of the rain-

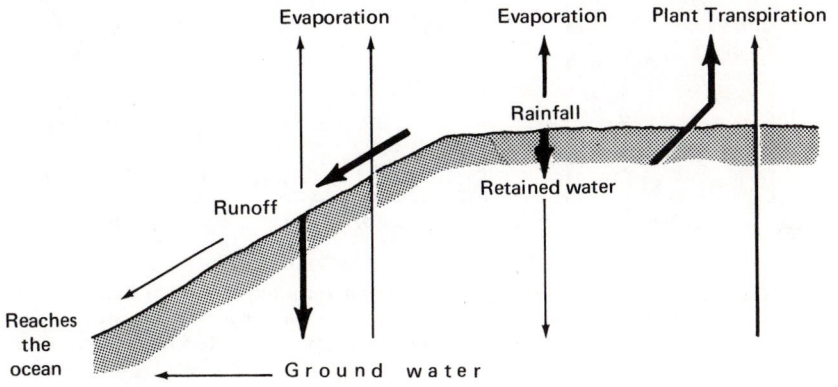

Fig. 42. Diagram showing the fate of rain in arid regions. The soil-retained water is important to plants. The run-off water seeps down to the ground water in the dry valleys and is only seldom reached by roots.

water runs off and a further portion evaporates (Fig. 42). How much of the water remains in the soil, and is thus available to the plants, is determined by the texture of the soil. In humid regions the sandy soils are dry because they retain only small amounts of rainwater whereas the clay soils are wet. The reverse is true for arid regions.

On flat ground in arid regions water does not sink down to great depths and thus does not reach the ground water. Only the upper soil layers are damp, and the depth to which the water penetrates depends upon the field capacity of the soil. Let us assume that 50 mm rain fall upon a dry desert soil and that it completely soaks into the ground. If the soil is sandy, then the upper 50 cm are wetted to field capacity. If the soil is a finely granulated clay with a field capacity 5 times as large, the water can only penetrate to a depth of 10 cm. On rocky ground with small cracks the water goes down much further, possibly 100 cm (Fig. 43).

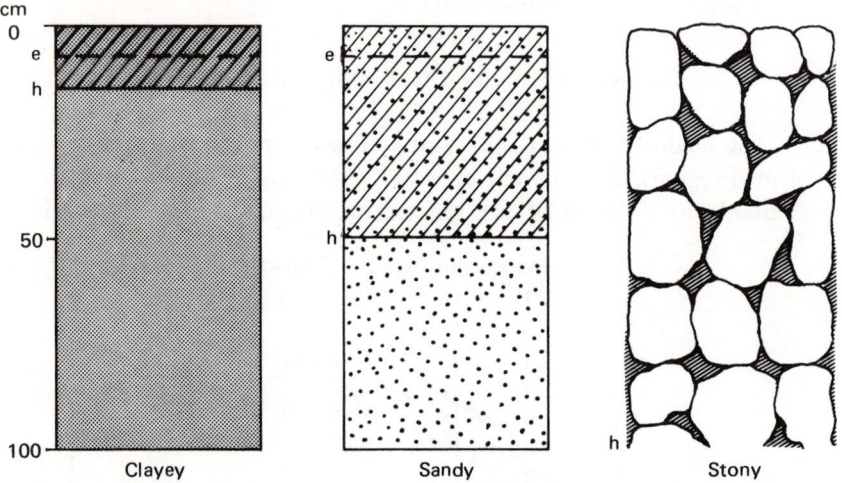

Fig. 43. Diagramatic representation of water retention in various kinds of soil following a rainfall of 50 mm in arid regions. h—h=lower level of moistened soil; e—e = lower level to which the soil dries out again. The clay soil retains 50% the sandy 90% and the stony 100%.

Evaporation follows upon the rain. If the upper 5 cm of a clay soil dry out, then 50 percent of the water originally entering the soil is lost. Sandy soils do not dry out so much, and even if the upper 5 cm were to dry out this would involve a loss of only 10 percent of the water. On stony ground there is almost no evaporation, and nearly all of the water is retained. This means that, *contrary to the situation in humid regions, the clay soils form the driest habitats whereas the sandy soils offer better water supplies. Fissured rocky ground provides the wettest habitat* if there is no run-off from the rock surface and if there is enough fine soil in the cracks to retain the water.

Such considerations have been confirmed by measurements in the Negev desert. Comparison of areas with an identical absolute rainfall revealed that loess soil offered the plants the equivalent of 35 mm rainfall in available water, rocky habitats with a relatively high run-off 50 mm, sandy soils 90 mm, and dry valleys with a large inflow the equivalent of 250—500 mm rainfall.

That the sandy soils are favorable habitats for plants in arid regions can be seen from the fact that the same type of vegetation occurs on sand at a lower rainfall than on clay. In the Sudan, *Acacia tortilis* is found on sandy soil in a zone which has a rainfall of 50 to 250 mm. On clay, however, it is found only in a zone with a rainfall of 400 mm. The *Acacia mellifera* savanna begins on sandy soil at a rainfall of 250—400 mm. In the short-grass prairie region of the Great Plains in western Nebraska a tall grass prairie occurs on sandy soil although it otherwise occurs farther east at a higher rainfall. The favorable water supply on stony or rocky ridges is marked in arid regions by the occurrence of trees in the midst of the lower-statured vegetation growing on the fine-grained soils.

If sandy soil or the soil in rocky clefts is wet down to the ground water, the roots of the plants also grow down this far, thus securing their water supply. The following example is worthy of mention.

North of Basra in Mesopotamia ground water is present at a depth of 15 m and is constantly replenished from the Tigris and Euphrates via gravel strata. Since, however, the rainfall merely amounts to 120 mm annually, the upper soil layer alone is damp and the plants are unable to reach down to the ground water. As a result there is only a sparse ephemeral vegetation following the winter rains. The native population has dug wells and uses the water for cultivating vegetables which they plant in furrows and irrigate several times daily, since the temperature can reach 50° C. The soil rapidly turns brackish due to the very high evaporation so that vegetables cannot be planted in the same spot for more than one year. Between the vegetable plants, however, *Tamarix articulata* cuttings are planted and rapidly take root. In the second year the furrows are not irrigated but the soil is still damp down to the ground water from the previous year. The tamarisk roots can therefore grow deeper and deeper in the ensuing years until they finally reach ground water, after which their water supply is secure and they can develop into large trees. They are cut down every 25 years for fuel and propagated by suckers from the stumps; in this way the desert farmland is transformed into tamarisk forest. *Deserts with deep-lying ground water can be converted into forest* if the soil is irrigated to such an extent in the first year after the trees have been planted that it is wetted down to the ground water.

7 Subtropical Arid Regions of the Different Floristic Realms

At the time when the conquest of the desert by plants took place, during the evolution of a terrestrial vegetation, the floristic realms were already differentiated. Since the plant families, or, speaking more generally, the taxa of the various floristic realms differ in their genetic constitution, adaptation to life under arid conditions has also taken different directions in the various floristic realms. Not only do the deserts differ floristically, not even the life forms are necessarily similar, although convergences do occur (p. 4).

a Holarctic Realm

The largest of the subtropical deserts is the northern Saharo-Arabian desert. It borders in the east immediately onto the Irano-Turanian and central Asiatic deserts, which have cold winters: The northern limit for productive date cultivation forms the border between the two. Chenopodiaceae are especially well represented in these deserts, partly on account of the extensive occurrence of saline soils. Succulent species of *Euphorbia* are found only in western Morocco; most of the species are xerophytic dwarf shrubs, some of them broom-like bushes. The only grasses present are xeromorphic with hard leaves: *Stipa tenacissima* and *Lygeum spartum* (transitional zone), *Panicum turgidum, Aristida pungens,* etc. Many ephemeral species appear after a good winter rain. Shrubs confined to wet habitats are *Tamarix, Nitraria,* and *Ziziphus.* Palaeotropic elements are numerous, including the species of *Acacia* found in the dry valleys carrying ground water.

In the U.S. only the southern Californian desert can be considered to be a subtropical Holarctic desert. The arid regions in northern Arizona, Utah, and Nevada already have a very cold winter.

b Palaeotropic Realm

Widely differing types of desert region belong to this floristic realm.

1. The southern Sahara, with the Sahel providing a transition zone to the summer-rain region of the Sudan. Grasses with less hard leaves *(Aristida, Eragrostis, Paniceae)* are much more common here. There are also far more shrubs *(Acacia, Commiphora, Maerua, Grewia)* as well as herbs *(Callotropis, Crotalaria, Aerva)*, etc., which are also typical of the Thar or Sind desert.

2. The Namib on the coast of Southwest Africa. The outer Namib to a width of 50 km is a brackish fog desert, usually rainless, although in 1934, when over 100 mm rain were recorded, there was a rich growth of ephemeral *Mesembryanthemum* spp. Otherwise, plants grow only in the dry valleys beneath which ground water is present, or on slopes receiving condensation water from the fog. Succulents are represented by *Euphorbia, Aloe, Asclepiadaceae*, etc. The well-known *Welwitschia mirabilis* grows on the border between the inner and outer Namib: It is a plant independent of ground water or fog moisture and grows only in the moist soil of the flatter drainage furrows. The inner Namib is a summer-rain region with *Aristida* grassland and shrubs *(Acacia, Commiphora, Tamarix)* and even trees in the larger dry valley. Toward the south the Namib gives way to the Namal, and here there is already a winter rainy season and the various species of *Mesembryanthemum* are more numerous.

3. The Karroo joins up with the Namaland in the east and stretches through North Capeland into the Orange Free State. The two rainy seasons favor the development of innumerable succulents: on rocky habitats the larger species of *Euphorbia, Portulacaria*, and *Cotyledon* as well as many smaller Crassulaceae, succulent Liliaceae, and *Mesembryanthemum* spp. on quartz veins (Fig. 44). The vast flat areas are covered with dwarf shrubs, mainly Compositae. Woody plants such as *Acacia, Rhus, Euclea, Olea, Diospyros*, and even *Salix capensis* grow in the dry valleys with ground water. In the tran-

Fig. 44. Great Karroo near Laingsburg (South Africa) with succulent *Euphorbia, Rhigozum obovatum, Rhus burchelli* and dwarf shrubs.

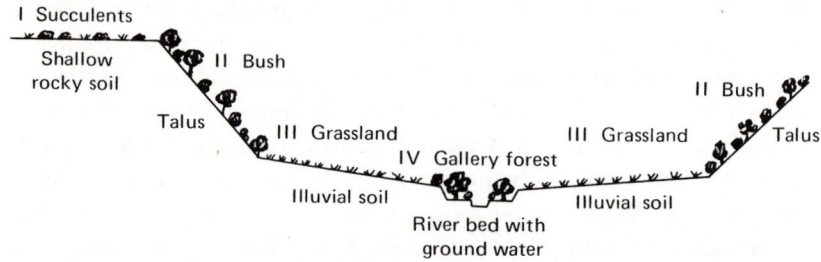

Fig. 45. Vegetation profile of a valley of the Upper Karroo near Fauresmith (South Africa). Distribution of the plant cover determined by differences in soil. Bush with *Oleo, Rhus* and *Euclea*.

sitional region of the upper Karroo the grassland of the summer-rain region as found growing on deep, fine-grained soils, whereas on shallow rocky ground the Karroo succulents still abound (Fig. 45).

4. Extensive arid regions also occur in the tropical parts of East Africa as well as in a small area in the rain shadow between the Pare- and West Usambara mountains where very odd succulents are to be found *(Adena globosa,* the boulder-like *Pyrenacantha, Euphorbia tirrucalli, Caralluma, Cissus quadrangularis, Sanseviera,* etc.). This is pro-

bably the driest region near the equator, with an annual mean temperature of 28° C and a rainfall of only 100—200 mm. In northern Kenya, western Ethiopia, Somaliland, and on Socotra there are even more extensive arid regions where *Adenium socotranum* (Apocynaceae), a plant with bizarre succulent stem achieving diameters of up to 2 m, is to be found. The baobab *(Adansonia digitata)* is distributed along the dry boundary of the tropical woodlands all over Africa. It is remarkable for its ugly shape and attains a height of 20 m. The amount of water stored in its thick, soft-wood trunk has been estimated in one case at 120,000 liters.

c Neotropic Realm

In this floristic realm, too, there are several semidesert regions:

1. The Sonoran desert (northern Mexico and southern Arizona), although acutally in North America, belongs to the Neotropic realm, floristically speaking. Extensive investigations of this desert (or rather semidesert) have been carried out from the Desert Laboratory in Tucson, Arizona. The vegetation, with its tall candelabra cacti, is

Fig. 46. Sonoran desert near Tucson, Arizona. Slope with giant cacti *(Carnegiea gigantea)*, erosion gulley with *Fouquieria splendens* and *Acacia* bushes; in immediate foreground, *Cercidium microphyllum*.

termed a "cactus forest" (Fig. 46). By means of a sort of bellows mechanism these succulents can store so much water that they are capable of surviving for more than a year without any further water uptake. They have a shallow root system, but within 24 hours following wetting of the upper soil layer, fine absorbing roots are put out and the water-storing tissue fills up. Apart from succulent cacti, other ecological types are represented here: Winter- and summer-ephemerals, poikilohydric ferns, malakophyllous half-shrubs *(Encelia)*, sclerophyllous species, stenohydric plants, and the deciduous *Fouquiera*, which develops new leaves after each heavy rain shower, although they afterwards rapidly turn yellow. Wide, flat, dry areas are covered with the particularly drought-resistant creosote bush *(Larrea divaricata)*, which smells strongly of creosote whenever its leaves are wetted. *Encelia* provides a good example of the widespread phenomenon that the degree to which the leaves are xeromorphic depends upon the water supply (cell-sap concentration), so that from habitat to habitat or during the course of the year all stages from large, green leaves to very small leaves, thickly covered with white, woolly hairs can be found.

In mountainous regions the cactus desert is succeeded in altitude first by a *Prosopis* shrub savanna with leaf-succulents *(Agave, Yucca, Dasylirion, Nolina)* and then by a sclerophyllous, evergreen oak forest.

2. In South America at the eastern foot and in the lee of the High Andes a *Larrea* desert strechtes more than 2,000 km from north Argentina as far as north Patagonia. The predominating species, *Larrea divaricata,* is probably identical with that to be found in Arizona.

3. The Peruvian-Chilean coastal desert is, at its most extreme, just as dry as the Namib, although the fog plays a greater role here because of the steepness of the coast in places. *Tillandsia* (Bromeliaceae), the only true fog plant known to exist among the flowering plants, occurs here. Although it cannot take up water from the humid air like the lichens, it does directly suck in the water drops from condensed fog with the aid of special leaf scales. The rosettes of these plants sit loosely on the sandy ground without developing roots.

The "Garua," as the fog blanket is termed in Peru, hangs at an altitude of 600 m for months on end during the cooler season. The soil on the slopes is so wet that a carpet of herbage grows, called

Loma vegetation, and is used for grazing. Although absent nowadays, woody plants formerly grew there. The quantity of water from fog condensations collected under the trees in a *Eucalyptus* plantation amounted to the equivalent of a rainfall of 600 mm. In north Chile the slopes exposed to fog are densely covered with cacti which are draped with lichens. Further south, near Frey Jorge, there is a true mist forest immediately adjoining a cactus semi-desert on the fog-free slopes. In the neighborhood of the large saltpeter deposits in north Chile the desert is completely barren, and only along the rivers fed by the snowfields of the High Andes is there vegetation or irrigated farmland.

4. Arid regions also occur in the equatorial zone of South America. These are the cactus thornbush semideserts of islands in the Caribbean Sea north of Venezuela and on the coasts exposed to the trade winds (see p. 31—32). The Caatinga, the dry region of northeast Brazil, can be included in this list.

d Australian Realm

The whole of central Australia is arid, despite which it has no climatic deserts. Sand dune regions (Gibson desert and Simpson desert), although not climatically the driest parts of Australia, are desert-like in character, as, too, are the Gibber Plains (bare, stony areas produced by over-grazing). The vegetation of the driest parts, with scanty rainfall at all times of year, is composed of salt bush *(Atriplex vesicaria)* and the blue bush *(Kochia sedifolia)*, both of them Chenopodiaceae. They occur either in pure populations or mixed. The soils upon which *Atriplex* grows contain little chloride (about 0.1 percent of the dry weight), but since these loamy soils dry out to a considerable extent, the concentration of chloride can in fact be very high. The cell-sap concentration is also correspondingly high (usually 40—50 atm, chlorides accounting for 60—70 percent of the total), and it is in fact an euhalophyte, the growth of which is enhanced by salt. A certain degree of salt excretion is achieved by means of the short-lived vesiculated hairs, which are continuously being replaced. *Atriplex* is a half-shrub, lives for about 12 years, and possesses, as do most halophytes, weakly succulent leaves and a root system spreading widely at a depth of about 10—20 cm above a chalk caliche layer. Therefore the bushes grow rather far apart.

In contrast, *Kochia sedifolia* is said to be long-lived. Its root system not only penetrates to a depth of 3—4 meters into the cracks in the caliche but also spreads equally far laterally. This species grows wherever rainwater percolates to greater depths, such as on a light or stony soil. The cell-sap concentration of *Kochia* is only half that of *Atriplex,* and the part played by chloride is also smaller (about 20—40 percent). *Kochia sedifolia* is thus probably a facultative halophyte. It can attain dominance if the climate becomes more humid.

In the salt-bush region there are scattered sand dunes or sandy areas where moisture conditions are more favorable and where the soil is not saline: shrubs such as *Acacia, Casuarina,* and *Eremophila* can be found. Tree-like species of *Heterodendron* and *Myoporum* as well as species of *Eremophila* and *Cassia* are confined to silty soils.

The most widely occurring species in central Australia is *Acacia aneura* ("mulga"). It dominates large areas which look like a gray sea when seen from the air. The shrub reaches a height of 4–6 m and has thin, cylindrical or somewhat flattened, resin-covered phyllodes. Its root system is well developed and penetrates the hard soil layers to a depth of about 2 m. Owing to the irregular rainfall (p. 99), flowering is not connected with any particular season but rather with the occurrence of rain. Fruits and seeds develop after a heavy rain shower, and at the same time the ground is carpeted with white, yellow, and pink everlasting plants belonging to the Compositae family (Fig. 47).

Acacia aneura is sensitive to salt but can survive long periods of drought. In dry habitats the bushes grow well apart, but in the wet depressions they form thickets. It is a hydrostable species with a slimy cell sap which, even after a long drought, has a concentration of only about 25 atm. The porcupine grasses *(Triodia, Plectrachne)* are another important group, collectively forming what is known as "spinifex" grassland. They are sclerophyllous species with very hard, rolled-up, pointed, perennial leaves with a resin covering, and they form large round cushions, or cupolas in the case of *Triodia pungens* with a height of up to 2 m.

Triodia basedowii dominates the sandy areas of the most arid part of western Australia. Its dense root system goes straight down for 3 m. Older cushions disintegrate and form garlands. Other characteristic genera, represented by many species, are *Eremophila, Dodo-*

naea, Hakea, Grevillea, etc. The structure of the vegetation is determined by the qualities of the soil and by the sheet floods following heavy rains, both factors leading to a complicated interweaving of various vegetational units. The increased summer rainfall in the north is accompanied by the appearance of savannas with evergreen species of *Eucalyptus* and a normally developed grass cover.

Fig. 47. Mulga vegetation in the interior of Australia near Wiluna following rain. Large bushes are *Acacia aneura,* smaller bushes *Eremophila* spec. Ground densely covered with temporarily active everlasting plants *(Waitzia aurea* and white *Helipterum* species).

The most outstanding characteristic of the vegetation in the dry areas of Australia is the lack of succulent species. An adaptation of this sort has not been achieved by the flora of Australia. Instead, the species are almost exclusively sclerophyllous, even in wet habitats, and are often covered with resin. This is especially conspicuous in the case of *Eremophila* species, the leaves of which are shiny but give a rather dry impression until they are picked and squeezed, when they readily lose a juice with a concentration of about 15 atm. Only among the halophytes are the same genera of Chenopodiaceae and Zygophyllaceae, and a few Mesembryanhemums found elsewhere in the world.

8 Types of Arid Landscape

In extremely arid regions plant cover is so sparse that the scenery is dominated entirely by its rocks. The following types of desert can be distinguished and will be given the names commonly used to describe similar areas in the Sahara:

1. *The rocky desert or Hamada* is mainly found on plateaus of the table mountains (mesas), from which all the finer products of weathering have been blown away and where the exposed rocks have

Fig. 48. Great "Fischfluss" canyon in the desert of southern S. W. Africa.

undergone severe wind erosion due to sand blast. The surface is covered with a pavement formed of hand-sized stones, darkly stained by desert-varnish (Mn-oxides) which lend a forbidding aspect to the landscape. Beneath the stony pavement there may be a water-repellant, dusty soil which is rich in gypsum and salt if it has originated in marine sediments, thus preventing the development of a plant cover. Hamada areas are cleft by deep erosion valleys with steep, rubble-covered slopes (Fig. 48). In the cracks and crevices of the rocks a few xero-halophytic species can take hold.

2. *The gravel desert or Serir (Reg)* arises from heterogeneous (e. g., conglomerate) parent rock. The cementing substance is readily weathered and removed by the wind, and the harder pebbles collect on the

surface. Such autochthonous gravel deserts are in contrast to the allochthonous ones, consisting of extensive alluvial deposits from which, again, the finer material has been blown away. Under the darkly stained gravel layer there may also be a crust, cemented hard with gypsum. It is a particularly monotonous type of desert, slightly undulating, with broad, shallow, sand-filled valleys offering better growth conditions for plants typical of sandy soil and for a few halophytes.

3. *The sandy desert, Erg or Areg,* has formed in the large basin areas by the deposition of sand blown off the raised ground. This also contributes to the formation of dunes: If there is a prevailing wind direction then sickle-shaped dunes are formed (barchanes), gently sloping on the convex, windward side, and with a steep slope on the concave, lee side. The dunes move in the wind direction, but if this varies periodically the crest of the dunes alters while the base remains fixed. A thin covering of iron oxide on the sand grains accounts for the bright red color of the dunes in hot, dry regions. Near the coast, where the air is more humid, the color changes to yellowish-brown.

These mobile and therefore barren dunes can store water because rain sinks in readily and does not evaporate. Even at an annual rainfall of only 100 mm a fresh ground-water horizon is present so that water can be obtained by sinking wells.

If the sandy covering is not very deep, colonization by plants is possible (nonhalophytes such as dune grasses, *Ziziphus,* etc.). Perennial species, including shrubs, serve as sand catchers and grow up through the sand that has accumulated around them, and then trap still more sand. In this manner each plant can form its own dune-hillock (several meters high), called a Nebka. These miniature dunes lend a characteristic note to the entire landscape.

4. *The dry valleys or Wadis (Oueds),* known in South Africa as riviers and in America as washes or arroyos, are an important feature in all deserts. They mostly originated in the past (Pluvial period), when the rainfall was higher. The dry valleys commence as scarcely noticeable erosion gulleys which then unite to form deep ditches or small valleys and often end in deep canyons. Gravel and sand are deposited by the water as it drains off after a shower. Some of the salt is washed out and the soil is soaked to a considerable depth, thus providing favorable conditions for the growth of nonhalophytic or halophytic plants such as tamarsk, as mentioned earlier in the

discussion of "contracted vegetation". The beds of the larger dry valleys bear no vegetation owing to the redistribution of the soil by occasional floods. Vegetation is confined to the valley sides which are safe from floods and the degree of luxuriance depends upon the amount of water held in the alluvial deposits. There is often a permanent underground flow of water, and in such cases dense, nonhalophytic woodlands are present.

5. *The pans, dayas, sebkhas or shotts* are hollows or larger depressions in which alluvial silt or clay particles are deposited. If there is subterranean drainage (in karst areas), they do not turn brackish. This is also true of the takyr, or delta-like formations at the valley exits, from which a part of the water drains off after a particularly heavy rainfall. The heavy soils, however, provide unfavorable habitats since the water can scarcely penetrate the soil and the ground rapidly dries out again after a flood. For this reason, mainly algae and ephemeral species grow on takyr soil. If there is no outflow and all the water evaporates, then salt concentration takes place and in such salt pans (sebkhas or shotts) compact layers of salt form in the deepest places. On the edges, where the salt concentration is lower, hygro-halophytes take hold. The salt content of the ground water is often low, and a salt crust forms only on the surface. If a thin layer of sand is deposited on the surface of such a salt pan, then there can be no capillary rise of water and hence no salt concentration. Plants soon establish themselves on the sandy deposits and serve to trap even more sand so that a hillocky or Nebka type of landscape is formed around the pans.

6. *Oases* are those sites in the desert with a dense vegetation, where water of a low salt concentration reaches the surface, either by means of normal springs or artesian wells. As has already been mentioned, hygrophilic species can grow here. Such oases are nowadays densely populated, and the natural vegetation has been replaced by cultivated plants or by weeds.

Oases with abundant water are often fringed by salt pans (shotts) where the excess water collects and evaporates.

9 Productivity of the Plant Cover in Arid Regions

It has been demonstrated that the primary production of the arid grassland of Southwest Africa rises linearly with rainfall and amounts to about 1,000 kg/ha dry mass per 100 mm mean rainfall (p. 87). Phytomass production is subject to certain variations in one and the same region. It is higher in a year with good rainfall, lower in a bad year. These variations, however, are not proportional to the absolute amount of rain and are somewhat damped by the fact that the species composition of the vegetation cannot change very rapidly: only the plants already existing alter their rates of development. The larger production in the areas with higher average rainfall is due to their supporting different, taller, and more productive grasses.

In the extreme desert near Cairo the production of ephemeral vegetation has been measured after the upper 25 cm of the soil had been soaked by winter rain amounting to 23.4 mm. Of this, 68 percent was lost unproductively by evaporation; transpiration of the ephemerals during the winter months accounted for 7.3 mm or 32 percent, which is the equivalent of 730 kg water per 100 m². Over the same area the ephemerals produced 9.384 kg fresh mass, or 0.518 kg dry substance. This gives a transpiration coefficienit of 730 : 0.518 = 1,409 which, compared with values for central European crops (400—600), is very high and is to be attributed to the low air humidity of the desert. If the production is calculated per 100 mm rain, then a value of something over 221 kg is obtained per hectare, i. e., less than a quarter of that of the more productive grass species in Southwest Africa growing under more favorable conditions.

A very large number of ecophysiological investigations have been carried out in desert regions, particularly concerning water economy. The most important result has been the discovery that xerophytes grow wherever the roots can reach soil water, thus rendering is possible for the plants to maintain a certain degree of photosynthesis and hence to produce organic matter. The most recent investigations in the Negev desert with poikilohydric lichens and some homoiohydric flowering plants and halophytes warrant mention (16). *Ephemeral species fulfill a buffer role* in that they exploit the excess water which the perennials are unable to utilize in a year of good rain. If, as in years of drought, there is no such surplus, the ephemerals simply do not develop. The phytomass of the perennials is determined by the rainfall in dry years, since these have somehow to be survived. If they have developed too prolifically after a series of good years, some of the perennials die or parts of the individual plants die off in the succeeding periods of drought, thus reducing the transpiring surface. This is the explanation of the observation that nearly all shrubs in arid regions have some dead branches and is the reason why young growth is seldom present. It is not a sign that the population is dying out, but rather an indication of the fact that regeneration can occur only if an old plant dies. The majority of dwarf woody plants achieve a great age, often of several hundred years. The oldest plants so far recognized are not the giant trees, *Sequoiadendron giganteum,* of mesic habitats in the Sierra Nevada of California but bristlecone pine *(Pinus longaeva,* = *P. aristata* p.p.*)*, a tree growing in the very arid mountainous regions of southeastern California, eastward to Nevada and Utah at altitudes above 3000 m. It may live even longer than 4,500 years, and such old stems have a ribbon-like trunk because the secondary growth took place in one direction only.

IV Sclerophyllous Vegetation of the Regions with Winter Rains

1 Transitional Zone Between Deserts and the Winter Rain Sclerophyllous Region

The boundary between true desert and semidesert, although not always clearly defined, is to be found in the zone where the increasing winter rains lead to the replacement of the "contracted vegetation" by a diffuse vegetation. About 25 percent of the total area in the semidesert is covered by vegetation, and the floristic composition of this plant cover varies just as greatly from one floristic realm to the other as is the case with the true deserts. North of the Sahara the malakophyllous *Artemisia herba-alba* and the sclerophyllous grasses *Stipa tenacissima* (halfa grass) and *Lygeum spartum* (esparto grass) are the most abundant species. Although *Artemisia* generally grows on heavy loess or loamy soil it has been found growing in Tunis in places where secondary $CaCO_3$ deposits are present at a depth of 10 cm. At a depth of 5—10 cm the soil is densely permeated by its roots, some of which even reach down 60 cm. *Stipa* prefers the high ground with a stony covering. A soil profile revealed the following: 2—5 cm stony pavement, beneath this 30 cm of loamy soil with dense root growth, below this a gravel layer with a hard upper crust which appeared to present an obstacle to the roots. Numerous roots originating at the base of the grass tuft spread far on all sides at a depth of 10—20 cm, so that although the tussocks themselves are 0,5—2 m apart, their root tips are, in fact, in contact with one another. Solitary individuals of *Arthrophytum* grow between the grasses. The soils are not saline. *Lygeum spartum*, on the other hand, is characteristic of soils containing gypsum (calcium sulphate) and even tolerates a certain amount of salt.

The halfa grass is cut and provides fibers for weaving, for the production of coarse ropes, and for paper manufacture. Whereas *Stipa tenacissima* is to be found only from southeastern Spain and eastern Morocco as far as Homs in Libya, *Artemisia herba-alba* is found in the Near East and has in many places replaced the original grassland, after overgrazing.

With a further increase in rainfall isolated trees occur, such as *Pistacia atlantica* in the west and *P. mutica* in the east, or *Juniperus phoenicea*. It is posible that the latter was originally always present among the halfa grasses until it was eliminated by man. Communities with scattered trees finally lead to the sclerophyllous woodlands.

In the transitional zone in California *Artemisia californica* and the half-shrubs *Salvia* and *Eriogonum* spp. (Polygonaceae) are found.

The transitional zone in northern Chile is a dwarf shrub semidesert with Compositae *(Haplopappus)*, columnar cacti, and *Puya* (large Bromeliaceae). This is succeeded by savanna with *Acacia caven*, the grass cover nowadays consisting of annual European grasses.

The so-called "renoster" formation *(Elytropappus rhinocerotis, Compositae)* in South Africa can be regarded as typical of a winter-rain region with low rainfall. In Australia, where there are no true deserts, the transitional zone is occupied by the mallee scrub consisting of shrubby species of *Eucalyptus*, the branches of which originate on an underground tuberous stem (ligno-tuber). Scattered *Eucalyptus* trees with an undergrowth of *Kochia sedoides* sometimes occur.

2 Mediterranean Sclerophyllous Zone

The climatic conditions prevailing in this zone can be seen in the diagrams in Fig. 49. Cyclonic rains occur in winter, while the hot, dry summer is a result of the Azores high pressure zone. Although there is no really cold season frosts do occasionlly occur, and the winter is cool enough for the main growing season to be the spring. Since some of our most ancient civilizations originated in the Mediterranean region the zonal vegetation was long ago forced to give way to cul-

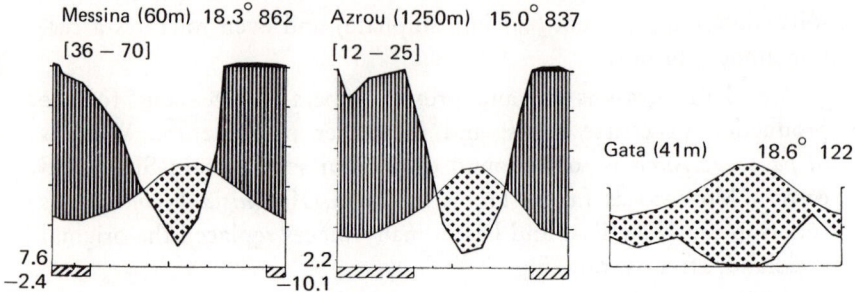

Fig. 49. Climatic diagrams from Messina, Sicily, Azrou in the montane belt of the Central Atlas (Marocco) and Cabo de Gata (S. E. Spain) = driest place in Europe (desert).

tivation. The slopes have been deforested and used for grazing, with resultant soil erosion, so that nowadays only varying stages of degradation remain. There is no doubt, nevertheless, that the original zonal vegetation was evergreen sclerophyllous forest with *Quercus ilex*. Small remnants of this asociation have provided the following data concerning the original forests:
Quercetum ilicis

Tree layer: 15—18 m tall closed canopy, composed exclusively of *Quercus ilex*.
Shrub layer: 3—5 m tall, *Buxus sempervirens, Viburnum tinus, Phillyrea media, Ph. angustifolia, Pistacia lentiscus, P. terebinthus, Rhamnus alaternus, Rosa sempervirens*, etc.; lianas—*Smilax, Lonicera* and *Clematis*.
Herb layer: Approximately 50 cm tall, sparse, *Ruscus aculeatus, Rubia peregrina, Asparagus acutifolius, Asplenium adiantum-nigrum, Carex distachya*, etc.
Moss layer: Very sparse.

A terra-rossa profile is found beneath these low forests, consisting of a litter layer, a dark humus horizon and, beneath this, a 1 to

2 m deep, clay-containing, plastic, bright red horizon. On cultivated land the upper layers are missing due to erosion, so that the red color is visible at the surface.

A change in the appearance of this region takes place in March, when many of the shrubs start to bloom. The height of their flowering season, as well as that of *Quercus ilex*, is in May, while *Rosa, Lonicera* and *Clematis* are still blooming in June. A relatively dormant period then follows as a result of the coincidence of the hottest and driest seasons. Growth only recommences with the autumn rains, which may then even lead to an additional flowering of the sclerophyllous trees.

Quercus ilex extends from the western Mediterranean region to the Peloponnesus and Euboea. Related species extend eastward to Afghanistan. In the west, *Quercus suber* is also to be found (not on limestone). These two species are replaced by *Quercus coccifera* in the eastern Mediterranean. The dominant species in the tree layer of the hot lower belts in Spain and North Africa are the wild olive tree *(Olea oleaster)*, *Ceratonia siliqua*, and *Pistacia lentiscus; Chamaerops humilis,* Europe's sole palm, also occurs here. In North Africa, from Morocco to Tunisia, the distribution of *Quercus ilex* is montane (see Fig. 50), above an intercalated coniferous belt consisting of *Tetraclinis (Callitris)* and *Pinus halepensis* (Aleppo pine). The southeast corner of Spain, with a rainfall of only 130 to 200 mm, is almost desert-like.

Nowadays there are only a few remaining places, in the mountanous regions of North Africa, where typical *Quercus ilex* forest still exists. Elsewhere the trees are cut down every 20 years, while still young, and they regenerate by means and shoots from the old stump. This leads to the formation of a maqui, consisting of bushes the height of a man. Maqui is also encountered on slopes where the soil is too shallow to support tall forest. Sclerophyllous species, usually shrub-like in form, may develop into big trees in a suitable habitat and can achieve a considerable age. Imposing old specimens of *Quercus ilex* can be seen in gardens and parks. In places where the young woody plants are cut every six to eight years and the areas regularly burned and grazed, the trees disappear entirely and open societies called garigue are formed (phrygana in Greece, tomillares in Spain, batha in Palestine).

Fig. 50. *Quercus ilex* forest above Azrou in the Central Atlas (Marocco). In the undergrowth *Rosa siculum, Lonicera etrusca.* etc.

These areas are often dominated by a single species such as the low cushions of *Quercus coccifera* or *Juniperus oxycedrus (Poterium spinosum* bushes, too, in the east) or *Cistus, Rosmarinus, Lavandula,* and *Thymus.* On limestone in the south of France the best grazing is provided by a *Brachypodium ramosum-Phlomis lychnitis* community. In springtime numerous therophytes (annuals) put in an appearance on otherwise bare spots as do such geophytes as *Iris,* Orchids *(Serapias, Ophrys)* and species of *Asphodelus.* An almost pure *Asphodelus* vegetation is all that finally remains in places seriously degraded by continuous fire and grazing. Although the garigue is a sea of flowers in spring it presents a severely scorched aspect in late summer. If cultivation or grazing is stopped then successions tending towards the true zonal vegetation take over, as shown in the scheme below (Fig. 50 a) for the south of France. On sandstone or on acid gravel the successions take a course similar to that on limestone, except that the individual stages are of a different floristic composition. Characteristic species are, for example, *Arbutus* and *Erica arborea.*

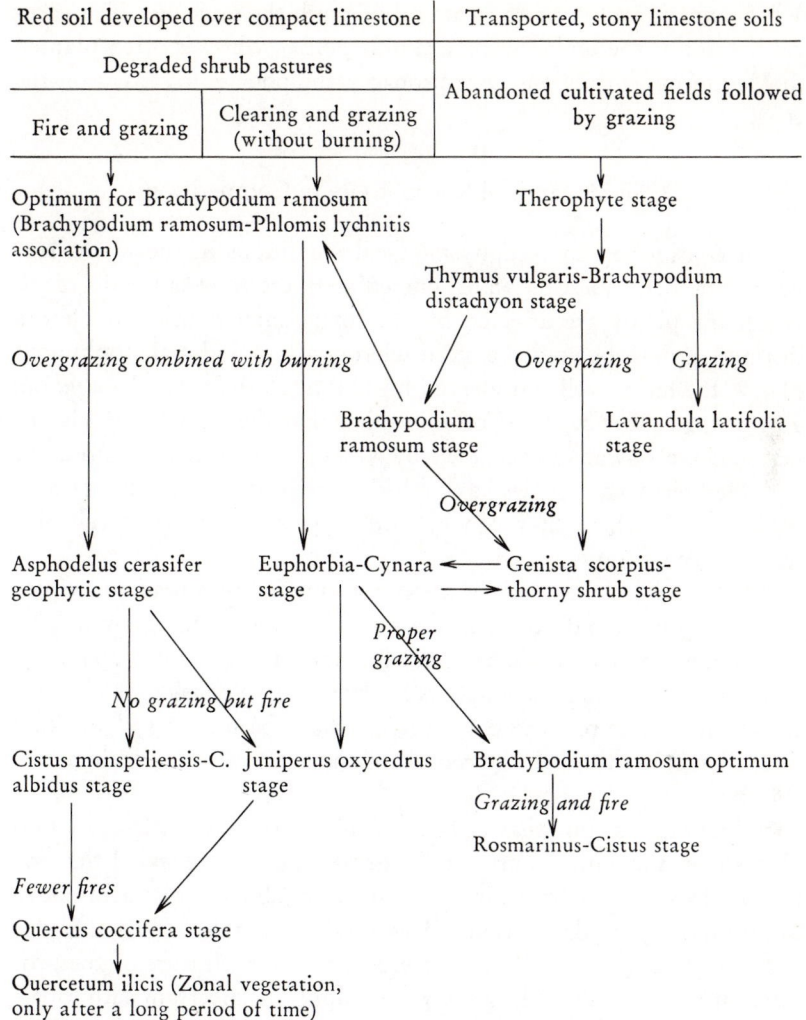

Red soil developed over compact limestone	Transported, stony limestone soils
Degraded shrub pastures	

Fire and grazing	Clearing and grazing (without burning)	Abandoned cultivated fields followed by grazing

Optimum for Brachypodium ramosum (Brachypodium ramosum-Phlomis lychnitis association)

Therophyte stage

Thymus vulgaris-Brachypodium distachyon stage

Overgrazing combined with burning

Overgrazing   *Grazing*

Brachypodium ramosum stage

Lavandula latifolia stage

Overgrazing

Asphodelus cerasifer geophytic stage

Euphorbia-Cynara stage ◄——— Genista scorpius- ——► thorny shrub stage

Proper grazing

No grazing but fire

Cistus monspeliensis-C. albidus stage

Juniperus oxycedrus stage

Brachypodium ramosum optimum

Grazing and fire

Rosmarinus-Cistus stage

Fewer fires

Quercus coccifera stage

Quercetum ilicis (Zonal vegetation, only after a long period of time)

Fig. 50 a. Plant succession under the influences of grazing and fire in southern France.

In the continental Mediterranean region of southern Anatolia *Pinus brutia* (related to *P. halepensis)* is widespread. It often constitutes the tree layer, below which sclerophyllous plants form a maqui. Since the pine is unable to regenerate in the maqui owing to lack of light, such woodlands can regenerate only after forest fires, which explains why the trees are all of much the same age. The natural habitat of the umbrella pine *(Pinus pinea)*, which is often planted in Mediterranean regions, was probably the poor sandy areas on the coast.

3 Significance of Sclerophylly in Competition

In considering the ecophysiological conditions in the Mediterranean region the question which immediately crops up is the degree to which the plants are affected by the long summer drought. Here a distinction must be made between sclerophyllous and malakophyllous plants, the latter well represented by *Cistus, Rosmarinus, Lavandula, Thymus,* etc. It should be borne in mind that the most favorable climatopes are nowadays occupied by vineyards or other cultures, the true Mediterranean species having been forced back into habitats with shallow soils where they can be said to grow under relatively unfavorable conditions.

If the underlying rock is deeply fissured, the abundant winter rains can penetrate deeply and the water is stored in the ground. This deep-lying water is available even in summer for plants capable of sending down roots through clefts in the rock to a considerable depth. In woody species roots 5 to 10 m length have been observed working their way down through the rock to the horizons which contain available water.

Observations on the cell-sap concentration of sclerophyllous plants over the entire course of the growing season revealed that the potential osmotic pressure increases only by about 4 to 5 atm above 21 atm during the dry season, which means that the water balance is not disturbed to any significant degree, and the hydrature of the protoplasm hardly falls. When the water supply is uncertain such a balance can only be maintained by a partial closure of the stomata and resultant limitation of gaseous exchange. Measurements of transpiration confirm that water losses in summer are three to six times greater in wet than in dry habitats. On the other hand, the cell-sap concentration of the poor individuals growing in extremely dry habitats reaches 30—50 atm. The climatopes which yield so much wine in autumn are far better provided with water, and a dormant summer period due to drought was certainly not a feature of the original sclerophyllous forests.

In contrast to the hydrostable sclerophyllous plants the malako-phyllous plants are highly labile in this respect. The cell-sap concentration of *Cistus, Thymus,* and *Viburnum tinus* can reach 40 atm in summer with an accompanying drastic reduction of the transpiring surface, achieved by shedding the greater part of the leaves, in some cases leaving only the buds. Such species do not root deeply. The laurel *(Laurus nobilis)* is not sclerophyllous and in the Mediterranean region its natural biotope is invariably in the shade or on northern slopes and nowadays it can only be found as forest in the altitudinal cloud belt on the Canary Islands, or in winter-rain regions with no pronounced summer drought, as, for example, in northern Anatolia.

The ecological significance of sclerophylly is thus to be seen in the ability of sclerophyllous species to conduct active gaseous exchange (400—500 stomata/mm^2) in the presence of an adequate water supply, but to cut it down radically by shutting the stomata when water is scarce. This enables them to survive mouths of drought with neither alteration in plasma hydrature nor reduction of leaf area. In autumn, when rains recommence, the plants immediately take up production again. Sclerophyllous plants are therefore able to compete successfully in the winter-rain regions not only with nonsclerophyllous evergreen species which are sensitive to drought (e. g., *Prunus laurocerasus),* but also with deciduous trees.

The situation changes at once, however, in the more humid winter-rain regions where the summer is not particularly dry or if the habitat, despite a typically Mediterranean climate, is itself perpetually wet, as is the case on northern slopes or in flood-plain forests. On mesic north slopes, the sclerophyllous species are replaced first by evergreen species like laurel, and then by deciduous trees. The deciduous oak, *Quercus pubescens,* with its larger production of organic material, replaces *Quercus ilex.*

Deciduous trees such as *Populus* and species of *Alnus, Ulmus campestris,* and *Platanus orientalis,* are found in the flood-plain forests of the Mediterranean region, and, in southwestern Anatolia, *Liquidambar orientalis,* a Tertiary relict, occurs. As soon as the point is reached at which the rivers dry up in the summer, however, deciduous woody species are no longer to be found. They are replaced by the evergreen oleander, *Nerium oleander.*

The productivity of plants depends largely upon their assimilation economy (Assimilat-Haushalt) and is the larger

1. the larger the proportion of the assimilated material which is used for increasing the productive leaf area,

2. the larger the ratio leaf area/leaf dry weight, i. e., the smaller the amount of material required to produce a given leaf area,

3. the greater the intensity of photosynthesis,

4. the longer the time over which the leaves can assimilate CO_2.

Exact values for the first point are not available, but it may be assumed that in deciduous species the contribution of leaf mass to the total phytomass is greater than in sclerophyllous species. As for the second point, the ratio is twice as large for the thin deciduous leaves as for the evergreen leaves. Measurements have shown that the intensity of photosynthesis per unit leaf area varies only slightly from deciduous to evergreen leaves. Regarding point four, evergreen leaves are of course at an advantage. This means that the deciduous species are superior in two considerations and the evergreens in only one.

Exact calculations have revealed that in the humid, mild climate on Lake Garda in Italy, where *Quercus ilex* and *Qu. pubescens* are to be found, the productivity in g/g dry branch weight was 22.9 for *Qu. pubescens* as compared with 17.9 for *Qu. ilex*, thus confirming the observation that deciduous species are able to compete successfully under these conditions of climate and habitat. On steep rock faces in the same climate, where a dry summer habitat results from runoff of the larger part of the rainwater, evergreen *Qu. ilex* bushes grow. In such biotopes *Qu. pubescens* is unable to compete.

4 Altitudinal Belts of the Mediterranean

In the mountainous regions of the Mediterranean a distinction must be made between

1. the humid altitudinal belts in the mountains on the northern margins of the Mediterranean zone in which, with increasing altitude, not only does the temperature decrease but at the same time the dry season disappears, and

2. the altitudinal belts with a summer drought, extending up to the alpine region.

In the first case the lowest evergreen sclerophyllous forest belt is succeeded by a sub-Mediterranean deciduous forest belt with oak *(Quercus pubescens)* and chestnut *(Castanea)*. Above this, at the summer cloud level, beech *(Fagus)* and fir *(Abies)* form a cloud forest. In the Apennines the timber line is formed by beech, which also occurs on Mount Etna and on Mount Olympus in Greece. In the Maritime Alps the beech belt is succeeded by a spruce *(Picea)* belt and in the Pyrenees by a belt of *Pinus sylvestris* and *P. uncinata.*

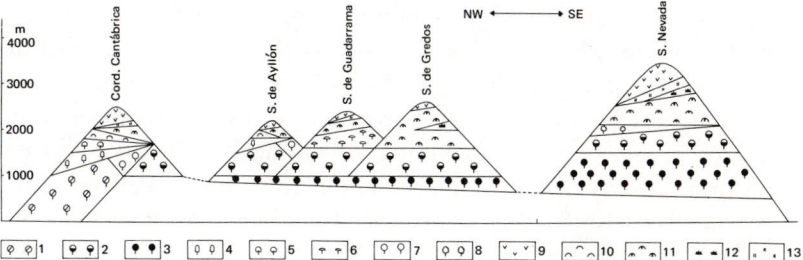

Fig. 51. Altitudinal belts in the high crystalline mountains of the Iberian peninsula shown in a N. W.-S. E. profile (from H. Ern). 1. Deciduous oak forest *(Qu. robur, Qu. petraea),* 2. *Qu. pyrenaica* forest, 3. *Qu. ilex* forest, 4. Beech forest *(Fagus sylvatica),* 5. Birch forest *(Betula verrucosa)* 6. Pine forest *(Pinus sylvestris),* 7. Mixed deciduous forest *(Quercus, Tilia, Acer),* 8. High altitude forest of the Sierra Nevada *(Sorbus, Prunus* etc.), 9. High alpine grass- and herbaceous-vegetation, 10. dwarf-shrub heath, *(Calluna, Vaccinium, Juniperus),* 11. Broom heath *(Cytisus, Genista, Erica),* 12. Thorn cushion belt, 13. *Festuca indigesta*-dry sward.

In the second case there is no deciduous forest belt. The Mediterranean sclerophyllous forest belt is followed immediately by a series of various coniferous forest belts. One finds, for example, on the southern slopes of the Taurus in Anatolia an upper Mediterranean belt with *Pinus brutia,* a weakly-developed montane belt with *Pinus nigra* ssp *pallasiana,* a high-montane belt with *Cedrus libanotica* and *Abies cilicica* (wetter) or species of *Juniperus* (drier) and a subalpine belt with *Juniperus excelsa* and *J. foetidissima.* But in the rainy northeastern corner of the Mediterranean in the Amanos mountains a cloud belt with *Fagus orientalis* is to be found. *Cedrus libanotica* occurs on Cyprus and a small relict is also present in Lebanon at 1,400—1,800 m above sea level. On Cyprus and Crete, as well as in Cyrenaica, cyprus always occurs in the upper Mediterranean belt in its natural form with horizontal branches. The frequently planted columnar form is a mutation. In the Atlas Mountains, from the eastern High Atlas to the Tunisian border, the subalpine belt at an altitude of more than 2,300 m above sea level consists of cedars *(Cedrus at-*

lantica), but the altitudinal belts vary greatly according to the course taken by the mountain ranges and the exposure of the slopes. Fig. 51 shows the complicated order of the altitudinal belts in the Spanish mountain ranges.

A difference in the altitudinal belts in arid and humid regions is recognizable even above the timber line in the alpine region. Whereas the situation in the humid mountain climate is similar to that in the Alps (chapter X), in the arid alpine regions the vegetation consists of thorny hemispherical cushion plants with many convergent species from different families. It is only possible to distinguish them from one another when they are flowering or fruiting. This belt is followed by a dry, grassy belt where hygrophilic plants (mostly endemic species related to arctic-alpine plants) are found on spots kept moist in summer by melting snow.

5 Mediterranean Steppes of High Plateaus

Anatolia has been chosen as an example of this type of country. It falls within the winter-rain region and has a central basin, 900 m above sea level, completely surrounded by mountains. These mountains catch a large portion of the winter rains, and in May the still wet but already warmed up ascending air masses lead to thunderstorms and a rainfall maximum (Fig. 7, Ankara). The total annual rainfall amounts to only 350 mm and there is a pronounced summer drought, while the months from December to March are cold (absolute minimum − 25° C), with occasional intervening thaws. No forest is capable of developing under such conditions and the pine forests of the encircling mountains (Mediterranean montane belt) are succeeded, via a shrub zone with *Juniperus, Quercus pubescens, Cistus laurifolius, Pirus elaeagrifolia* and *Colutea, Crataegus* and *Amygdalus* (dwarf almond) species by steppe, largely given over to arable land (winter wheat cultivation as dry farming) or to intensive grazing. This has resulted in degradation to an *Artemisia fragrans-Poa bulbosa* semidesert with many spring therophytes and geophytes. At greater altitudes abundant species of *Astragalus* (Tragacantha) and *Acantholimon* (Plumbaginaceae), which are especially characteristic of the cold Armenian and Iranian highland, form thorny hemispherical cushions. Originally, this area was herbaceous grass steppe reminiscent of the East European steppes (p. 169), except for the Mediterranean floristic elements. The soil shows a typical chernozem (p. 167) profile, although the A-horizon is not very rich in humus. The cold winter, and the summer drought account for the brief growth period of only four months, whereby the occurrence of the rainfall maximum in May is of great significance.

The most favorable season here is the spring. The first geophytes are already flowering by February and March *(Crocus, Ornithogalum, Gagea,* etc.) followed, on overgrazed areas, by numerous small therophytes which, since they only root in the upper 20 cm of the soil, have disappeared again by June. The genuine perennial steppe species are fully developed by May and do not dry out until July. Their cell-sap concentration is low (10 to 15 atm) because the soil contains sufficient water in the spring, and it rises only shortly before the plants die off. A whole series of species, including the thorny cushions, bloom during the main drought season. They are equipped with long tap roots enabling them to take up water from the deeper soil horizons which are still moist in the summer: A root length of 7.65 m has been recorded in a 30-month-old specimen of *Alhagi.* Their cell-sap concentration, too, is below 15 atm.

The periphery of the Mediterranean was settled very early by man and may be looked upon as the cradle of human civilization. The Hittites of Anatolia were among these early settlers, as well as the inhabitants of the fertile crescent formed by the mountainous slopes bordering Mesopotamia to the west, north and east. The oldest traces of grain-growing have been found in the neighborhood of Jericho, Beidha, and Jarmo. Such steppe land provided suitable conditions not only for grain-growing but also for the support of cattle, while the surrounding forests offered both game and fuel. The inhabitants of the ancient settlements completely ruined the natural vegetation in the intervening thousands of years and in places which once were fertile country there is now desert. Soil erosion has set in and badlands with no sign of plant life are frequently encountered.

6 The Sclerophyllous Vegetation of the Winter-Rain Region of California

In contrast to the east-west extension of the winter-rain region in Europe which, thanks to the influence of the Mediterranean Sea reaches far into Asia and is still pronounced in the south Crimea, in Trans-Caucasia, in Iran and even in Afghanistan, the winter-rain region in western North America is limited by mountain ranges (Cascades and Sierra Nevada) to a narrow strip on the Pacific Coast. This strip extends down the West Coast from British Columbia to Lower (Baja) California, although in the north the rainfall is so high and the summer drought so brief that the forests are hygrophilic to mesophilic coniferous forests, rich in species.

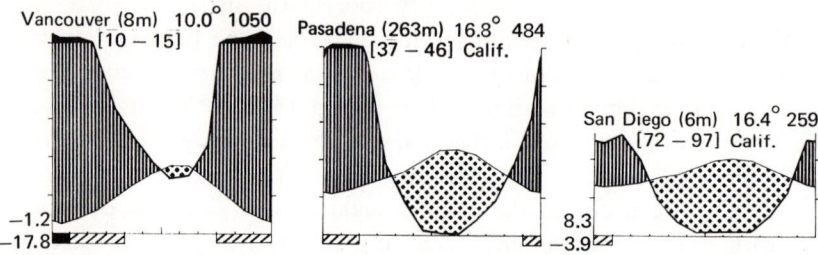

Fig. 52. Climatic diagrams of stations on the Pacific coast of N. America (from north to south) in the coniferous forest-region, sclerophyllous forest-region and the region transitional to the desert.

Central and southern California together form a sclerophyllous region, but Lower California is too arid (Fig. 52). The north-south gradient of rainfall explains why only in the northern part of the sclerophyllous region do evergreen oak forests occur, sometimes even mixed deciduous species, whereas in the south a shrub formation known as chaparral predominates, corresponding to the Mediterranean maqui. Since the present-day flora of westernmost America is quite similar to that of the Pliocene Epoch apparently little impoverishment took place in the Pleistocene Age and therefore the plant communities are very rich in species. Such genera as *Quercus* and *Cupressus* are represented by a large number of species, and many other genera entirely absent in Europe are to be found; for example, the important shrub genera *Ceanothus* (Rhamnaceae), represented by 40 species and *Arctostaphylos* represented by 45. Oone of the main species is the Rosaceae *Adenostoma fasciculatum* ("Chamise") with needle-like leaves; its distribution fairly exactly reflects the extent of the sclerophyllous zone. The strip of land immediately bordering the ocean north of latitude 36° N does not belong to the sclerophyllous zone. The fog resulting from the cool ocean currents renders the summers cool and wet so that hygrophilic, northern tree species are found.

Unlike the maqui, the chaparral is the natural zonal vegetation corresponding to a relatively low winter rainfall of 500 mm. Fires are common in this region, and were a natural factor even before the advent of man. The statistics of the U. S. Forest Service reveal that fire caused by lightning is very common in the chaparral region, rendering constant fire-watching necessary during thunderstorms. If a fire occurs every 12 years the character of the chaparral remains unchanged since the shrubs can sprout afresh. But if no fire occurs for a great lenght of time then species such as *Prunus ilicifolia* and *Rhamnus crocea* infiltrate. If one fire follows another within two years the seedlings of those shrubs which themselves cannot sprout afresh after a fire are killed off, and these woody species are eliminated.

The roots of sclerophyllous species reach far down into the ground because the upper soil layers are usually completely dried out in summer. By means of their roots, which may even penetrate from 4 to 8.5 m into the rocky fissures, the plants are able to obtain a certain amount of water in the dry season.

Availability of water can be recognized from the observation that within three weeks of a fire in the height of summer the shrubs begin to sprout again. After the loss of the transpiring surface a small amount of water suffices for the buds to open. The autumn rains have no direct effect since it takes a month for the water to sink to a depth of one meter. In the meantime, the temperature falls so much that growth does not commence. In April, when the water supply is good and the temperature rising, growth is at a maximum. The old evergreen leaves assimilate into the spring and are shed in June, by which time the new leaves can take over their function. All chaparral species possess a mycorrhiza and the *Ceanothus* species have nodules which assimilate atmospheric nitrogen.

Evergreen sclerophyllous forests are also found as a montane belt in North America, above the cactus desert in the mountains of south and central Arizona, at an altitude of 1200—1900 m. This is known as the encinal belt, and on the basis of the distribution of the different species of *Quercus* it can be subdivided into an upper and lower belt. The upper belt is succeeded altitudinally by a *Pinus ponderosa* belt. The chaparral genera *(Arbutus, Arctostaphylos, Ceanothus)* form the shrub layer in the encinal forest. Although Arizona has two rainy seasons the vegetation is strongly reminiscent of that of California, except that the sclerophyllous forests in the mountains are much better developed and are still original. East of the Sierra Nevada, in the state of Nevada, the winter rains fall off to about 150–250 mm, and at an altitude of 1,300 m the cold season lasts six to seven months. The vegetation is an *Artemisia tridentata* semidesert known as sagebrush. It occupies enormous areas in Nevada, Utah and the bordering states and is the cold-climate equivalent of the southern *Coleogyne* and *Larrea* semidesert. *Artemisia* prefers the heavy soils of the basins and gives way to Pinyon on elevated stony ground. Pinyon consists of low scattered tree communities with *Pinus monophylla* or *P. edulis* and *Juniperus* spp, including some cold-resistant chaparral species. True coniferous forests of *Pinus flexilis* and *P. aristata* commence at an altitude of about 2,000 m in the mountains, while further east *Pinus ponderosa* is found. Higher up this is replaced by *Pseudotsuga* and *Abies concolor*, while the timber line above 3,000 m is made up of *Picea englemannii* and *Abies lasiocarpa*. The dry southern slopes are often devoid of trees and *Artemisia* extends up to the alpine region. But the order of the altitudinal belts can vary greatly. The aspen *(Populus tremuloides)* is also frequently encountered where ground water or snowdrifts provide extrazonally mesic conditions.

Artemisia tridentata is a half-shrub, 1.5 to 2 m high and attains an age of 25 to 50 or more years. Its tap root extends about 3 m into the ground and gives off horizontal lateral roots extending far on all sides. In spring, after the snow has melted and the water supply is good, the cell-sap concentration is very low (about 10 to 15 atm). It soon rises to about 20 to 35 atm and can even amount to 70 atm in summer if there is an acute water shortage. At this point, like all malakophyllus plants, it sheds it leaves. The sagebrush semidesert is confined to the arid brown, semidesert, salt-free type of soil, with *Artemisia tridentata* as the dominant species, frequently accompanied by the dwarf-shrubs *Chrysothamnus* (Compositae) and *Tetradymia canescens* which are physiognomically similar.

Depressions with no outflow are invariably brackish in such an arid climate. The salt pans and salt lakes in these regions are the relicts of much larger Pleistocene lakes such as Lake Bonneville in Utah, the surface of which was 310 m above the Great Salt Lake and the extensive surrounding barren salt desert. Lake Bonneville covered an area of 32,000 km², with a maximum length of 586 km and a maximum width of 233 km. The present salt desert is more than 161 km long and 80 km wide. In 1906 the Great Salt Lake measured 120 km by 56 km when at its highest level. Its contours vary greatly, the average depth being a little above 5 m. Its salt content varies between 13.7 percent and 27.7 percent when it is saturated. About 80 percent of the salt is NaCl and the remainder consists of $MgCl_2$, Na_2SO_4, K_2SO_4, etc. (17). Halophytes flourish around the saline areas and the vegetation forms very distinct zones:

Hygro-halophytic *Allenrolfea* and *Salicornia* grow on the edge of the salt crust, followed by *Suaeda* and *Distichlis*. Further broad zones are occupied by *Sarcobatus vermiculatus* which requires ground water, and the xero-halophytic species *Atriplex confertifolia*. Species of *Kochia* and *Eurotia lanata* then lead on to a nonhalophytic zone with *Artemisia tridentata*. With the exception of the salt-excreting grass *(Distichlis)* all of the halophytes are Chenopodiaceae.

The climate in Utah is very similar to that of Ankara. The marked predominance of *Artemisia* is the result of overgrazing; such grasses as species af *Agropyron, Stipa* and *Fustuca* were formerly widespread.

7 The Chilean Sclerophyllous Region

Chile, lying at the western foot of the High Andes, is a long, narrow strip of land 200 km wide and 4,800 km long, extending from 18° S to 57° S. It exhibits every possible transition in vegetation from the rainless subtropical desert in the north, via sclerophyllous region, to the very wet, temperate and subarctic forests in the south. Winter rains are prevalent throughout (Fig. 53). The cold Humboldt current flowing along the entire coast modifies the summer drought so that temperatures are lower compared with those of California. The mean annual temperature of Pasadena at 34° N is 16.8° C whereas in Santiago, at 33° S it is only 13.9° C. Since Chile belongs to the neotropical floristic realm its flora is quite different from that of the Mediterranean region and of California. Only the cultivated areas offer a similar appearance since the same species are cultivated on the farms and in the gardens in all three regions. The sclerophyllous region occupies the central part of Chile and adjoins the arid regions in the north. It is represented only by remnants forming woodlands

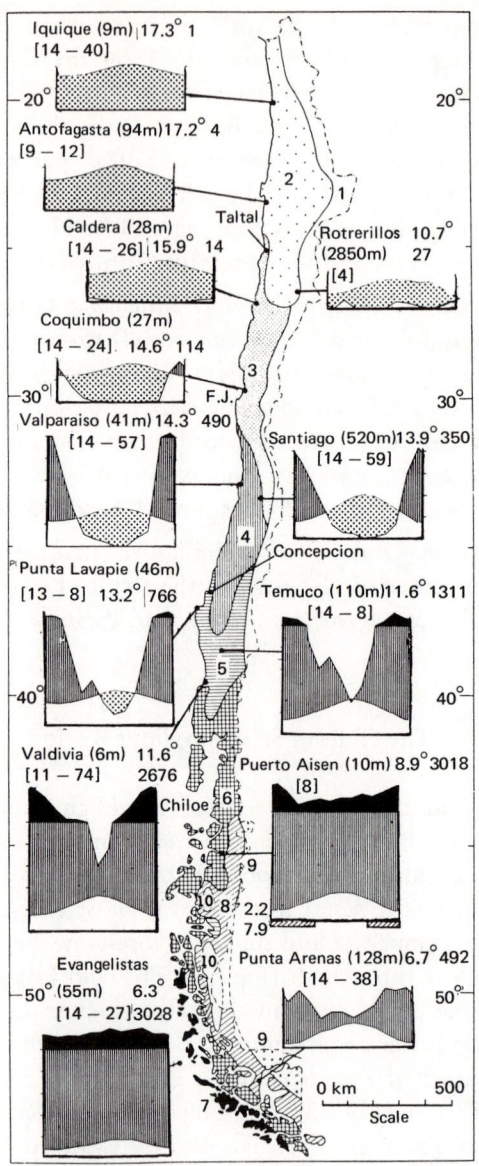

Fig. 53. Vegetation of Chile (modified from Schmithüsen) and climatic diagrams. N-Chile: 1 northern High Andes, 2 desert region, 3 dwarf shrub- and xerophytic shrub-region, 4 sclerophyllous region, 5 deciduous forest. S. Chile: 6 evergreen rainforests of the temperate zone, 7 tundra-like vegetation of the cold zone, 8 subantarctic deciduous forest, 9 Patagonian steppe, 10 southern Andes.

10 to 15 m tall, with such xerophytic species as *Lithraea caustica* (Anacardiaceae) which causes rashes and fever if touched, *Quillaja saponaria* (Rosaceae), *Peumus boldus* (Monimiaceae) or the Lauraceae *Cryptocarya* and *Beilschmiedia,* which have a preference for wet

Fig. 54. Landscape near Santiago (Chile). In the foreground, on rocky soil, flowering *Trichocereus.* In the valley, remnants of sclerophyllous vegetation. The grasses are adventitious annual Mediterranean species (Avena, etc).

ravines, in addition to a whole series of shrubby species. The endemic palm *Jubaea spectabilis* grows in a narrowly limited area northeast of Valparaiso; columnar cacti *(Trichocereus,* Fig. 54), and the large *Puya* spp. (Bromeliaceae) as well as the thorny Rhamnaceae, *Colletia* and *Prevoa,* are found in dry, rocky habitats.

On the Chilean side, the Andes drop steeply: Aconcagua, at 7,000 m, is only about 100 km from the coast. Talus communities predominate and altitudinal belts are difficult to recognize. The sclerophyllous vegetation extends up to about 1,400 m, and shrub communities lead on to the alpine region, with the occasional appearance of the conifer *Austrocedrus (Libocedrus) chilensis*. Alpine talus plants such as *Tropaeolum* spp, *Schizanthus* (a Solanaceae with zygomorphic flowers), as well as Amaryllidaceae *(Alstroemeria, Hippeastrum)* and species of *Calceolaria* are common. Flat, cushion-like plants are characteristic of the upper alpine belt *(Azorella* and other Umbelliferae).

The species occurring in these altitudinal belts, as well as to the south of the sclerophyllous zone, are already antarctic elements, as are the tree form *Nothofagus* species. Immediately south of Concepción, with decreasing summer drought, *Nothofagus obliqua* forest is found. The trees lose their leaves in the cool winter months. Further south, with a rainfall exceeding 2000 to 3000 mm, this forest is replaced by evergreen Valdivian temperate rain forests which are scarcely less luxuriant than the tropical rain forests, while their woody mass is probably even greater. The woody species are partly neotropical in origin. Also bamboos (Chusquea) are common. Part are antarctic elements as, for example, the evergreen *Nothofagus dombei*. A large number of ancient coniferous forms are found, especially in montane situations. Besides *Austrocedrus* and *Podocarpus* species, *Saxegothea, Fitzroya, Araucaria araucana (= A.imbricata), and Pilgerodendron uviferum* also deserve mention. The climate is very wet and cool but frost-free, so that the evergreen forest gives way to the so-called Magellan forests which stretch almost to the tip of South America, gradually becoming poorer in species and decreasing in height until finally they are only 6 to 8 m tall. The westerly islands are covered by bogs with cushion-like plants *(Sphagnum* is not important), a vegetation closely related floristically to that of the antarctic islands. Similar antarctic elements are found in New Zealand and in the mountains of Tasmania, indicating that these regions were formerly directly connected with each other via the antarctic continent. The bogs can be considered to be antarctic tundra.

8 The Vegetation of the South African Winter-Rain Region

Although confined to the outermost southwestern tip of Africa the South African winter-rain region comprises an entire floristic realm, the Cape. This small region is extraordinarily rich in species: In the Jonkershoek nature reserve alone 2,000 species have been recorded on 2,000 ha and an equal number on the 50 km stretch from Table Mountain to the Cape of Good Hope. Six hundred species of the genus *Erica*, 108 species of *Cliffortia* (Rosaceae), 115 species of *Muraltia* (Polygalaceae), 117 species of *Restio* (Restionaceae), and about 100 species of *Protea* are included. Particularly abundant among the sclerophyllous plants are the Proteaceae, a family which is otherwise well represented only in Australia, albeit by a different

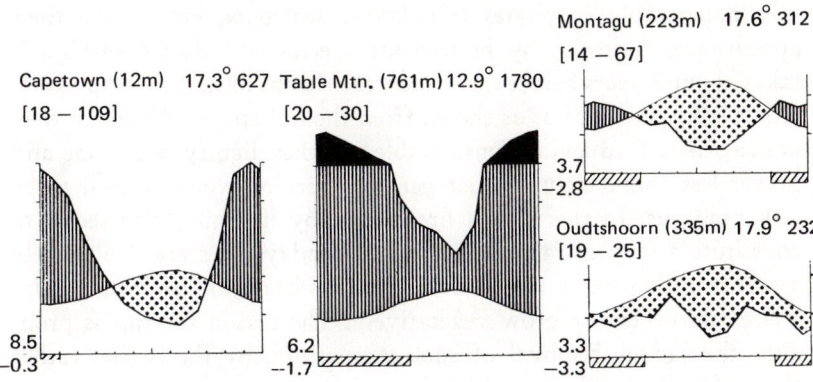

Fig. 55. Climatic diagrams from South Africa. Typical sclerophyllous region, damp montane climate (misty), transitional region and typical Karroo.

subfamily. A few genera also occur in South America. Many house plants of temperate latitudes orginally came from the Cape *(Pelargonium, Zantedeschia = Calla, Amaryllis, Clivia,* etc.). The climatic diagrams for Cape Town and Tangiers are comparable, except that the annual rainfall of the former is 260 mm less although the summer is slightly less dry (Fig. 55). The sclerophyllous vegetation of the Cape is known as Fynbos and is a proteaceous scrub, 1 to 4 m high, very similar to the maquis. The only tree species, *Leucadendron argenteum* (silver tree), is of very limited distribution and is confined to the humid slopes of Table Mountains, at altitudes below 500 m. In wet ravines tree stands can be found, but these are in fact the last outposts of the wet temperate forests of the southeast coast of Africa.

The leaves of the Proteaceae are sometimes very large, and although they have very little mechanical tissue they are nevertheless sclerophyllous because of their thick cuticle. As in all sclerophyllous plants the water balance of the Proteaceae shrubs is in a state of equilibrium, which means that the cell-sap concentration undergoes only very slight variations during the course of the year. The deeper layers of the soil, into which the roots penetrate, apparently contain water even in the summer. Cape soils are acid and poor in nutrients, which particularly suits both Proteaceae and Ericaceae. Fire is the most important ecological factor: In the year immediately after a fire innumerable geophytes *(Gladiolus, Watsonia,* etc.) make their appearance, followed by herbaceous species and dwarf shrubs. It takes about 7 years before the Proteaceae shrubs have grown up again, either from seedlings or as shoots from the old crown. Although they are capable of achieving considerable age they lignify with time and flower less abundantly, so that periodic burning would appear to be advantageous. In this region fire caused by lightning also seems to constitute a natural factor, although nowadays fires are deliberately set. It is interesting to note that bulb plants begin to flower only after a fire and otherwise grow vegetatively. The reason for this is probably the sudden removal of root competition by the bushes rather than a fertilizing effect resulting from the ash.

Rainfall increases with increasing altitude only on the southeastern slopes of the mountains, up which the warm, humid air from the Indian Ocean is forced to rise. The rainfall recorded at the Table Mountain station is three times as high as that of Cape Town, 750 m lower down. The Capeland is mountainous, with basins scattered between the mountain ranges which often wear a "table-cloth", or a covering of cloud rising up the southeastern slope and dispersing again on the northwestern slope. This leads to a wet mist on the high plateaus of the table mountains, and there is a tendency toward heathland *(Restio, Erica)* and even to the formation of moors (mossy mats of *Drosera* and species of *Utricularia*). Such succulents as *Rochea coccinea* grow between dry boulders.

Moving inland the rainfall decreases, particularly in the rain shadow of the mountain ranges. For this reason the sclerophyllous vegetation recedes further and further up the seaward slopes. In the rain-shadow area the dry form of Cape vegetation is met with, known as the renoster bush *(Elytropappus rhinocerotis,* Compositae).

This is succeeded inland by the semidesert vegetation of the Karroo (p. 103).

Since the colonization of the Cape in 1400 the sclerophyllous vegetation or Fynbos has spread extensively. Formerly an evergreen temperate forest with paleotropical elements stretched along the entire coast of southeast Africa beyond the southern tip of the continent (Cape Ahulhas). This has been cleared, however, apart from the area around Knysna, and its place has been taken by the sclerophyllous vegetation which reaches as far as Port Elizabeth, except in the cultivated areas.

9 The Vegetation of the Winter-Rain Regions of Australia

Perth in Southwestern Australia lies at approximately the same latitude as Cape Town and has a very similar climate (Fig. 56). However, winter rain falls not only on the southwestern corner of this continent, but also on the area around Adelaide in South Australia.

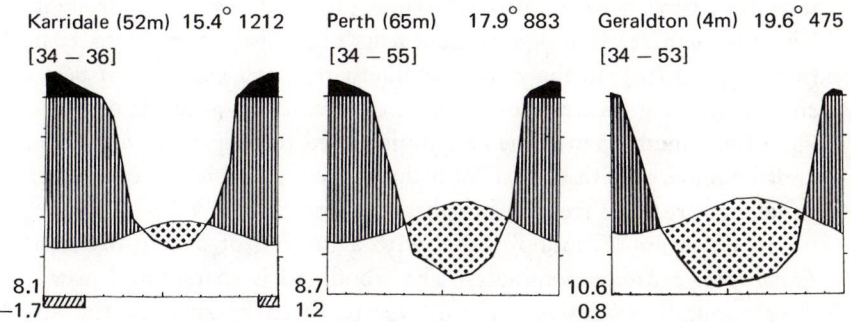

Fig. 56. Climatic diagrams from S. W. Australia. Stations in the Karri forest, in the Jarra forest, and in the shrub-heath (see also Fig. 59, Adelaide).

On account of the peculiar floristic situation (p. 4) the sclerophyllous vegetation differs in character from that in other winter-rain regions of the Earth. The tree form *(Eucalyptus* spp.) dominates and Proteaceae constitute a lower shrub layer or may even achieve predominancy on the sandy heaths. The leaves of *Eucalyptus* are leathery rather than hard. Peculiar to Australia are the "grass trees" *(Xanthorrhoea, Kingia)*, the Cycadaceae *Macrozamia*, and species of *Casuarina*.

Epacridaceae take the place of Ericaceae. Just as on the Cape, the soils are acid and poor in nutrients, containing SiO_2 and iron concretions which are the lateritic crusts of an earlier geological age when the climate was tropical. The parent rocks are among the oldest geological formations of the Earth. Indicative of the poverty of the soil is the fact that 47 species of *Drosera* (sundew) occur in the herbaceous layer of the forest around Perth. Wherever it is wet enough bracken *(Pteridium)* is to be found.

Rainfall increases to the south of Perth (up to 1,500 mm), but decreases to the north and inland. Each change in climate means that different species of *Eucalyptus* rise to dominance. The wetter the climate the taller the trees and therefore the greater the leaf area per hectare. Because the leaves hang vertically plenty of light can penetrate the space around the trunks so that the shrub layer is usually well developed, insofar as it has not been reduced by frequent fires.

"Jarrah" forest, which is completely dominated by *Eucalyptus marginata,* is characteristic of the climate corresponding to a Mediterranean type, with 650 to 1,250 mm rain and a summer drought. The trees can reach an age of 200 years and a height of 15 to 20 m (maximum 40 m). In the wetter, southern regions *Eucalyptus diversicolor* forms the "Karri" forest, with trees reaching a height of 60 to 75 m (maximum 85 m). The canopy is closed to 65 percent coverage, and the undergrowth consists of a shrub layer and a herbaceous layer developed from the fronds of the bracken (Fig. 57).

In the drier "Wandoo" zone, with a rainfall of 500 to 625 mm, *Eucalyptus redunca* dominates. The woodland is sparser and nowadays is almost exclusively given over to sheep grazing. In the absence of suitable indigenous grasses *Lolium rigidum* and the Mediterranean clover *Trifolium subterraneum,* which buries its fruits underground and provides a source of nitrogen, are planted. Owing to the poverty of the soil fertilization with superphosphate is essential before sowing; in view of the large areas involved both of these processes are carried out from the air.

In the zone receiving 300 to 500 mm rain annually many species of *Eucalyptus* are found scattered over the landscape. Nowadays, however, this is the winter-wheat zone with farms of several hundreds of hectares, completely mechanized and run by 2 or 3 men. Rust diseases render wheat cultivation in wetter zones uneconomical.

When the mean annual rainfall drops below 300 mm *Eucalyptus* disappears and the extensive grazing lands provided by the bush semi-desert commence (p. 108).

In South Australia there is no humid winter-rain region, but otherwise the situation is similar to that in southwestern Australia,

Fig. 57. *Eucalyptus diversicolor* forest in S. W. Australia. Undergrowth of *Acacia pulchella* and bracken *(Pteridium esculentum)*.

only rather more complicated because mixed forest communities occur, each consisting of several species of *Eucalyptus*. Furthermore, the region is mountainous, which again leads to a marked differentiation of the vegetation.

Apart from the forests already described there are vast areas of heathland with 0.5 to 1 m tall Proteaceae. The bushes are capable of growing on the poorest of sands, and even the least demanding of the *Eucalyptus* species are unable to compete with them. Such areas are not cultivated and are hardly used for grazing. It is all the more remarkable that so very many species grow on such an impoverished sandy soil. On 100 m², 90 species have been recorded, including 63 woody species, mainly Proteaceae or Myrtaceae. *Drosera* species and a tuberous *Utricularia* also occur.

Thorough ecophysiological investigations have been carried out on such a heath in South Australia at a rainfall of 450 mm annually and with 7 months of drought in summer. Soil temperatures at a depth of 15 cm varied between 4.1° C and 36.0° C and at 30 cm between 5.8° C and 29° C. The dominating sclerophyllous species are the bushy *Eucalyptus bacteri*, 9 Proteaceae, 2 *Casuarina* species, *Xanthorrhoea*, Leguminosae, etc.

The main growth season is the dry summer since the soil remains wet to a great depth. Smaller perennial species (42 percent) root only in the upper 30 to 60 cm, and grow in the spring. *Drosera* and orchids are ephemeral species and root at a depth of only 5 to 7 cm. It is interesting to note that the water in sandy soil with a wilting point of 0.7 to 1 percent is very unevenly distributed because the larger species drain the rainfall toward their stems where it enters the soil. The composition of the heathland is conditioned by fire. Immediately after a fire the grass tree *Xanthorrhoea* begins to sprout, and in fact it blooms only after a fire. *Banksia* (Protaceae) regenerates by means of seedlings and its share in the above-ground phytomass rises to 15 percent by the 15th year. The larger part of the dry substances of 25-year-old specimens is accounted for by the large fruits which open only after a fire.

Banksia is one of the many *pyrophytes* very commonly encountered in Australia. *These plants are capable of regeneration only after fire* because the woody fruits do not otherwise open, which would suggest that fires caused by lightning have always played a natural role in Australia. Forest and heath are often burned down nowadays because the woody plants are of no commercial value and hinder grazing. In the farmer's opinion "one blade of grass is worth more than two trees". Many Proteaceae and Myrtaceae as well as the coniferous *Actinostrobus* are pyrophytes and even *Eucalyptus* species seed more prolifically after a fire.

The nutrients in a heath that has not been burned for a long time are bound up in the fruits of *Banksia*, in the old leaves of

Xanthorrhoea and in the accumulating litter. A fifty-year-old community degenerates on this account and not until a fire has caused mineralization of the nutrients can a new succession be initiated.

Eucalyptus marginata woodland is ecophysiologically typical of a sclerophyllous vegetation. The roots penetrate the hard lateritic crust in places to a depth of more than 2 m. There is no summer dormant period and the water balance is maintained by a partial closing down of the stomata from 10 a. m. to 3 p. m., with a resultant decrease in transpiration. The cell-sap concentration in winter was found to be 16.3 atm and is probably only slightly higher in summer (18).

V The Warm-Temperate Vegetational Zone

1 Winter-rain Regions with No Summer Drought and No Cold Season

In connection with the sclerophyllous region it has already been mentioned that there are forests which, although still in the winter-rain region, are not exposed to a pronounced drought. Summer rains are plentiful despite the occurrence of a rainfall minimum at this time of year. This, together with a mild winter and the absence of a cold season means that the climate may be termed warm temperate. The redwood *(Sequoia sempervirens)* forests of California, north of San Francisco, represent a hygrophilic type of warm-temperate forest. This species attains a height of 100 m and is thus taller than the giant big tree *Sequoiadendron* which grows only in the Sierra Nevada of California. Whereas the trunks of the former remain slender and the trees only achieve an age of 500 years (maximum 1,800), 4,000 annual rings have been counted in big trees. Further to the north *Tsuga heterophyllia, Thuja plicata* and *Pseudotsuga menziesii* are the most frequently encountered species (Fig. 58).

The valdivian forest of southern Chile is a warm-temperate rain forest, but in Africa there is no corresponding humid winter-rain region since the continent does not extend further than 35° S. In the furthermost southwestern corner of Australia the Karri forest represents the humid winter-rain vegetation. In the Mediterranean the

Fig. 58. Damp, oceanic coniferous forest with *Pseudotsuga menziesii*, *Tsuga heterophylla* and *Thuja plicata* on the Hoh River (Olympic Nat. Park). Cf. climatic diagram of Vancouver (Fig. 52).

warm temperate insubrien region has a rainfall maximum in summer, and in the Colchis region on the north Anatolian Black Sea coast the rainy winters are so mild and the summers so wet that tea can be cultivated. The southwestern coast of the Caspian Sea also has a similar type of Climate (Talysh).

However, the true warm-temperate regions are to be found on the eastern coasts of the continents, with summer rains.

2 The Warm-Temperate Forest Zones on the Eastern Coasts of the Continents

Because the eastern seaboards of the continents are exposed to trade- or monsoon-winds, their rainfall maximum occurs in the warm season. In Southeast Asia, eastern Brazil and eastern Australia the wet tropical forests are replaced, as the latitude increases, by subtropical forests which finally give way to warm-temperate forests. It is extremely difficult to draw a line between these three zones. The forests are evergreen, and the temperature steadily decreases. The land rises from the coast and orographic rains account for the absence of a dry season; the mean leaf size gradually decreases, and the floristic composition of the forests alters. Pronouncedly tropical species become scarcer, but the tree ferns, which prefer a cool, damp climate, are more common. Even at the lower temperatures, however, conditions remain favorable for a forest vegetation. These forests have not been subjected to ecophysiological investigations because the decisive factor under such favorable conditions is competition, and this is a factor difficult to elucidate.

A peculiar situation exists in Africa because northeastern Africa borders the Red Sea and not the Indian Ocean. The area has a desert-like climate. Only at the upper end of the steep side of the high plateau is the desert-like character somewhat modified by sparse rains and a humidifying cloud belt caused by ascending air. In the tropical parts of East Africa the climate is relatively dry because of the southwest and northeasterly monsoons blowing across the land masses. In the south the foreland of the Drakensberg Mountains and the steep slopes of the high land mass to their north are more humid. Here, too, the forests are subtropical to warm temperate in character, especially those at the southernmost tip of the continent.

In North America this zone is not particularly well defined either, because of the cold air masses which at intervals move down from the north as far as the Gulf of Mexico, as well as to extensive marshy regions and sandy plains in the southeastern lowlands which hinder the development of a zonal vegetation. Nevertheless, near the coast in Louisiana, Florida, and Georgia, as far north as North Carolina, evergreen tree species typical of warm-temperate regions are found.

3 Eucalyptus-Nothofagus Forests of Southeastern Australia and Tasmania

The wet tropical-subtropical evergreen forests on Australia's east coast consist mainly of Indomalayan elements foreign to the Australian realm. They extend as far as southern New South Wales on rich, usually volcanic soils. Only in southern Victoria and in Tasmania does the Australian element dominate with the genus *Eucalyptus*, combined, however, with some antarctic elements, the most important of which are *Nothofagus cunninghamii* and the tree fern *Dick-*

sonia antarctica, as well as a series of other species in Tasmania. *Eucalyptus regnans* can attain a height of 110 m (the earlier figure of 145 m could not be verified) in this wet climate where a cold season is lacking (Fig. 59). The composition of the forest depends upon the frequency of fire.

1. In the wettest parts of western Tasmania where no fires occur a tree stratum of *Nothofagus* and *Atherosperma moschata* (Monimiaceae) develops to a height of 40 m and, below it, a 3 m-high stratum of the fern *Dicksonia* which is able to grow even when receiving only 1 percent of the total light. Hymenophyllaceae and mosses abound as epiphytes.

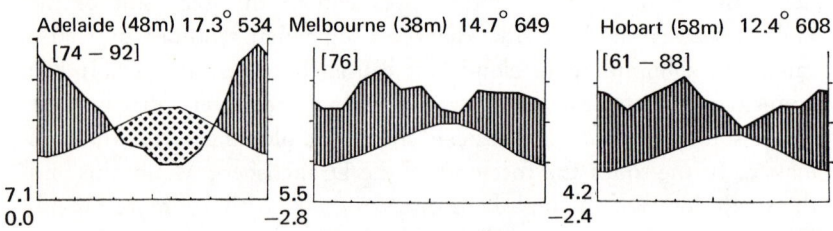

Fig. 59. Climatic diagrams from the sclerophyllous region of S. Australia and the warm temperate regions of Victoria and Tasmania.

2. If forest fires occur more frequently than every 350 years, mixed forest develops, consisting of three strata. Apart from the two strata mentioned above there is a further loftier stratum consisting of the three largest species of Eucalyptus (75 to 90 m). That the trees in this stratum are all of one age is an indication that germination of their seedlings took place simultaneously over an extensive area after a fire. Although the *Eucalytpus* and *Nothofagus* strata are destroyed in the fire, the fruits can still open and the undamaged seeds are dispersed and germinate. The more rapidly growing *Eucalyptus* overtakes *Nothofagus,* with the result that two tree strata are formed. Although the tree-ferns lose their leaves in a fire, they are able to develop new ones at the tip of the stem. Regeneration of *Eucalyptus* below *Nothofagus* is impossible owing to lack of light and can take place only after another fire.

3. If forest fires occur once or twice in a century, then *Nothofagus* is replaced by other more rapidly growing but lower-statured tree species such as *Pomaderris, Olearia, Acacia,* etc.

4. A pure, low *Eucalytpus* vegetation results where fire occurs every 10 to 20 years.

5. Still more frequent fires lead to a degradation of the forests. An open moor results, with "button grass" *(Mesomelaena sphaero-cephala* (Cyperaceae), with scattered Myrtaceae shrubs together with *Drosera, Utricularia,* and Restionaceae.

4 Warm-Temperate Forests of New Zealand

New Zealand's forests warrant special mention. Although both islands are relatively near to the Australian continent and were probably directly connected with it in geological past, this connection must have been interrupted before the flora of the Australian realm was fully developed. There is not a single native species of *Eucalyptus* or *Acacia* in New Zealand, and the Proteaceae are represented by only two species.

In the north of North Island there are still subtropical forests consisting of coniferous *Agathis australis* and palms, and along the coast there are even Mangroves with low *Avicennia* bushes. The forest species are melanesic elements of the Palaeotropic realm.

Forests of this type occur even on South Island although its climate is definitely temperate, despite the absence of a cold winter season in the lower-lying country. The coniferous genera *Podocarpus* and *Dacrydium,* which are distributed throughout the entire Southern Hemisphere, are very common here. The antarctic element, represented by five evergreen *Nothofagus* species, plays an important role in the forests of both North and South Islands. These mutually exclusive forest species form a mosaic for which there is no satisfactory ecological or climatological explanation. The plant cover gives the impression of not being in a state of equilibrium with its present-day environment. Rather than being differentiated by soils or other ecological factors, it seems that the vegetation reflects mostly historical factors. Seventeen hundred years ago North Island was covered by a thick layer of volcanic ash. The first pioneers were Podocarpaceae disseminated by birds; they were then gradually replaced by forests containing tropical elements as well as by *Nothofagus* forest in some of the mountainous regions. In the Pleistocene Age, South Island was covered by large glaciers so that here, too, the process of recolonization is still going on, since *Nothofagus* spreads very slowly.

In the extremely humid Fjord Country of the southwest, where the rainfall exceeds 6,000 mm, the *Nothofagus* forests are similar in nature to those of southern Chile. A peculiarity is, however, represented by the bare strips, 2 to 6 m wide, suggestive of avalanches, found in the middle of the forest on steep slopes. When the weight of the tree layer becomes too great for the rocky slopes, then the entire vegetation, inclusive of roots and soil, slides down because of the force of gravity. The naked rock which remains is then recolonized by lichens, mosses, and ferns, followed by shrubs and finally by trees, until the process repeats itself.

The imported European red deer presents a great danger to the forests of New Zealand where originally the only mammals were bats. It is impossible to control the multiplication of the deer, and they hinder the regeneration of the often inaccessible *Nothofagus* forests, so that the danger of erosion and flooding is greatly increased. The Australian opossum (Kuzu), also imported, is equally dangerous. It has confined itself to a tree species growing at the timber line in the high mountains. It strips the trees of their leaves completely, thus bringing about their death and increasing the danger of soil erosion on the steep slopes.

New Zealand provides an example of the extreme danger of disturbing the natural equilibrium by introducing new plants or animals. The damage done is often irreparable.

VI Nemoral Zone or Deciduous Forest Zone in Temperate Climates

1 Leaf-Shedding as an Adaptation to the Cold Winter

A temperate climatic zone with a marked but not too prolonged cold season occurs only in the Northern Hemisphere. Apart from certain mountainous districts in the southern Andes and in New Zealand, it is absent from the Southern Hemisphere. The phenomenon of facultative leaf-shedding has already been met with in the tropics. There the leaves turn yellow and are shed only when the water balance is disturbed by a lengthy period of drought, and this leaf loss decreases water losses (p. 65). Leaf-shedding in the temperate zone, however, is an adaptation to the cold season. It is not facultative but obligatory, occurring even if the trees grow in a greenhouse where they are protected from the cold of winter. The factor responsible for setting off the change in color of the leaves in autumn, even before the first frost occurs, is unknown. It might well be the decreasing day length. A remarkable fact is that the various species of trees turn color within a very short period of time in central Europe, according to the phenological calendar between the 10th and 20th October), with no sharp distinction between places in the west and in the east. An evergreen broad leaf is neither resistant to cold nor to winter drought, that is to say, to sustained temperatures below freezing. In a central European climate *Prunus laurocerasus* (cherry laurel) invariably freezes in severe winters, and even a light frost causes the leaves to excrete CO_2 by day, which means that, although respiration continues, photosynthesis is blocked. *Ilex aquifolium* (holly) is atlantic in its distribution and *Hedera helix* (ivy) subatlantic, both species thus avoiding the eastern continental regions with their cold winters. The same holds true of the broom species, *Ulex* and *Sarothamnus*. The alpine rose *(Rhododendron)* and cranberry *(Vaccinium vitis-idea)* can survive the cold only beneath a covering of snow.

The loss of the thin deciduous leaves in winter and the protection of the buds from water losses represents a saving of material as compared with the freezing of the thick evergreen leaves. It is, however, essential that the leaves newly formed in spring then have a sufficiently long, warm summer period of at least four months to produce enough organic material for the growth and maturation of the lignified axial organs and for the formation of reserves for the fruits and for the buds of the following year. Even in their bare winter condition the twigs lose water, the extent varying from species to species. For this reason the central European beech avoids the zone with a cold east European winter, although the oak extends as far as the Urals. In extreme continental Siberia the only broad-leaved deciduous trees are the small-leaved birch *(Betula)*, the aspen *(Populus tremula)*, and the mountain ash *(Sorbus aucuparia)*, with small pinnate leaves. Where the summers are too cool and too short, evergreen conifers replace the deciduous species. Their xeromorphic needles are more resistant to cold in winter, and when warmer weather returns in spring they are capable of starting up photosynthesis immediately. In this way the short vegetational season can better be exploited. Whereas deciduous trees require a vegetational season lasting at least 120 days with a mean daily temperature above 10° C, the conifers manage with 30 days. However, the resistance of the latter varies from species to species. The yew *(Taxus)* does not extend east of the border of central Europe, like the ivy *(Hedera)*. The Scots pine *(Pinus sylvestris)* and the spruce *(Picea abies)* are very resistant, and *(Pinus sibirica (= Pinus cembra)* survive in Siberia, but the deciduous, needleleaved larch *(Larix sibirica)* extends furthest of all into the continental arctic regions, where it exploits the short summers in a state of high productivity. Whether the species with deciduous or those with evergreen assimilatory organs are more successful in competition and rise to dominance thus depends upon external conditions and the ecophysiological characteristics of the species themselves.

2 Distribution of the Temperate Deciduous Forests

From what has already been said it is clear that deciduous trees favor a climate with a warm vegetational season of 4—6 months with adequate rainfall, and a mild winter lasting 3—4 months. For this reason they avoid the extremely maritime as well as the extremely continental regions and favor what is termed the nemoral zone. In the Northern Hemisphere a climate of this kind, with the rainfall maximum occurring in summer, is to be found in eastern North America and in Eastern Asia, between the warm temperate and the cold- or arid-temperate zones. It is also found in western and central Europe north of the Mediterranean zone where, as a result of the influence of the Gulf Stream, winter rains are replaced by evenly distributed rainfall or by rainfall with a summer maximum and where the cold season is relatively short.

The Mediterranean winter-rain region with sclerophyllous vegetation extends widely from west to east, while to the north it is replaced by several differing vegetational zones. In the maritime region on the Atlantic coast to the northwest of Gibraltar, elements of evergreen warm-temperate laurel forests occur, with the evergreen species *Prunus lusitanica* (related to *P. laurocerasus)* and *Rhododendron ponticum* ssp. *baeticum.* Also to be found here are *Quercus lusitanica* ssp. *canariensis (Qu. mirbeckii); Drosophyllum lusitanicum,* an interesting insectivorous plant; and the tiny *Utricularia lusitanica,* besides the epiphytic fern *Davallia canaricum*. This type of vegetation, however, is rapidly succeeded by Atlantic heaths which extend in the coastal region as far as Scandinavia, to be replaced in the north by birch forests. Genuine laurel forests exist only on the humid, windward side of the Canary Islands (Tenerife). Further to the east a sub-Mediterranean zone is intercalated between the Mediterranean and nemoral zones. Although there are still winter rains, the summer

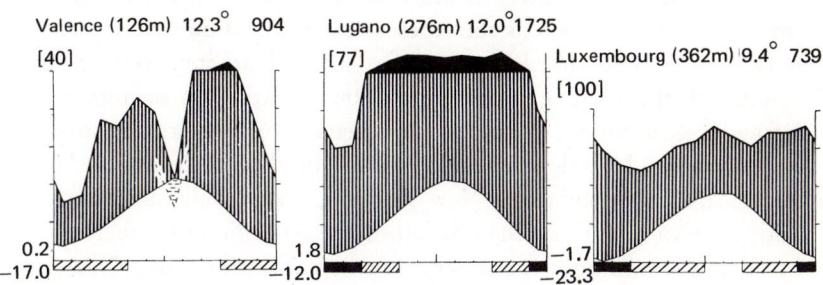

Fig. 60. Climatic diagrams from the sub-Mediterranean zone (still no cold season), from the warm and the damp deciduous forest zones as well as from central-European beech forest zone (see also Fig. 7, Hohenheim with rainfall maximum in summer).

drought is no longer pronounced and frost occurs regularly in all months of the winter (Fig. 60). Apart from one woody evergreen species *(Buxus)*, all of the tree species in this zone are deciduous e. g., *Quercus pubescens, Fraxinus ornus, Acer monspessulanum, Ostrya carpinifolia,* or the frequently cultivated chestnut *(Castanea sativa).* Consequently, this region cannot be termed a Mediterranean zone, but should preferably be considered as an intermediate zone. To the northeast of this sub-Mediterranean zone is the steppe zone, which only further north is replaced by various kinds of forest. In the Middle East, the Mediterranean sclerophyllous zone is succeeded by the Mediterranean steppes and semideserts.

3 The Atlantic Heath Regions

The Atlantic heaths represent stages in the degradation of deciduous forests, a process which has been going on since prehistoric times. Destruction is nowadays so complete that the heaths have long been considered to be the true zonal vegetation. The fact that the soils are extremely poor and acid and capable of supporting only a weak kind of heath vegetation was formerly attributed to leaching as a natural consequence of the humid climate. What has already been said with regard to tropical rainforests (p. 41) can also be applied here. As long as the natural forest vegetation remains untouched, leaching of nutrients from the biogeocenose does not occur and the reserves of nutrients are mainly stored in the aboveground phytomass. But as soon as the forest is cleared and burned, the larger part of the mineralized nutrients is lost and an impoverished soil results. If the ensuing heath vegetation is exploited or repeatedly burned, reforestation, in any case problematic, is rendered impossible. In uncolonized extreme oceanic regions in the southwest of South America, in Tasmania and New Zealand, with similar temperatures and twice to four times the rainfall of the Atlantic heath regions, we have seen *Nothofagus* forests growing in undisturbed luxuriance and with no sign whatever of degradation due to leaching of nutrients.

It is not easy to reconstruct the original composition of the west European forests, but in all probability oak was the most abundant species *(Qu. petraea* and *Qu. robur)*, together with birch *(Betula)* in the north, as well as the evergreen species *Ilex aquifolium*. Heath, as an independent community, occurred only on shallow or peaty soils in the clearings and otherwise formed the forest undergrowth. Only after destruction of the forests did it take possession of the entire area.

In the southern part of the coastal zone, species of broom dominate *(Ulex, Sarothamnus,* and *Genista* spp), accompanied by various species of *Erica*. In the central districts the broom species become scarce, leaving *Ulex europaeus, Sarothamnus scoparius,* and *Genista anglica* as the most important representatives. At the same time the numbers of Ericaceae greatly increase, above all *Calluna vulgaris* besides *Erica cinerea* and *E. tetralix*. *Empetrum, Vaccinium, Phyllodoce,* and *Cassiope* dominate in the north.

In the Tertiary Age oceanic coniferous species which still occur in the northwest of North America were found in Europe, but became extinct in the Ice Age.

One-quarter to one-third of Scotland is covered by *Calluna* heath, which is regularly burned. The iron podsol soils often have a cemented B-horizon forming hardpan ("Ortstein"). *Calluna vulgaris* is absolutely dominant. It is a dwarf shrub achieving a height of about 50 cm. It develops a dense web of roots in the upper 10 cm of soil, with a few roots going down 75—80 cm to the hardpan. Its very small leaves are sessile on short twigs, of which the larger portion is shed in autumn, thus reducing the danger from winter-drought in the cold season. The annual litter production of a dense community amounts to 421 kg/ha. If the heath is burned every 30 years, the development of the vegetation falls into three phases, each lasting ten years:

1. The reconstructive phase following the fire. Part of the nutrients is bound in the litter.

2. The phase of maturity, during which litter production increases and the rate of increase in phytomass drops.

3. The phase of degeneration, during which litter production remains constant but its destruction increases until a state of equilibrium is attained. After 35 years the standing phytomass amounts to 24,000 kg/ha and the litter to 17,000 kg/ha.

As a rule the heath is burned again after 8—15 years, without the phase of degeneration having been reached. In climates as humid as that of Scotland, fires are caused only by man. Natural fires caused by lightning hardly ever occurred in the original forests, so that their degradation has only taken place where man has intervened. All of the transitional forms from heath to moor are to be found. Four stages associated with increasing moisture are listed, the species being arranged in the order of their decreasing abundance:

1. *Erica cinerea, Calluna vulgaris, Deschampsia flexuosa, Vaccineum myrtillus*

2. *Calluna vulgaris, Erica tetralix, Juncus squarrosus*

3. *Erica tetralix, Molinia coerulea, Nardus stricta, Calluna vulgaris, Narthecium ossifragum*

4. *Erica tetralix, Trichophorum caespitosum, Eriophorum vaginatum, Myrica gale, Carex echinata.*

In Scotland the heath is utilized for grouse shooting and also for extensive sheep grazing, with 1.2—2.8 hectares per animal. In Germany on the "Lüneburger Heide," which is also purely anthropogenic in origin, farming was formerly common (buckwheat cultivation). The upper 10 cm of the raw humus layer were dug out in squares, used as stable litter, and returned to the fields as manure. This process hindered reforestation. Nowadays the heath area is no longer cultivated, and forests have grown up from birch and pine seeds blow in by the wind. In some places the heath is systematically reforested.

Apart from heaths, bogs are of frequent occurrence in extreme maritime regions. The climate is oceanic with small fluctuations of temperature: In Ireland, for example, the January temperatures are 3.5—3.7° C and those for July 14—16° C. Frosts may occur, but snow covers the ground only 3—10 days of the year. Rainfall amounts to 350—1,000 mm annually and is evenly distributed, varying from year to year by 25 percent at the most. Owing to the prevalence of cloudiness, the amount of sunshine is only 31 percent of the possible maximum. Under such conditions the danger of bog formation after deforestation is very great. Since a low, herbaceous vegetation loses less water due to transpiration than does the tree stratum of the forest, a rise in ground-water level is noticeable after deforestation in humid regions. This, in turn, favors the growth of peat-mosses, principally *Sphagnum* spp., although *Rhacomitrium lanuginosum*, too, is widespread. In regions where it rains on more than 235 days of the year, bogs may cover the entire area. Such "blanket bogs" are found in West Ireland, Wales, and Scotland, where the largest encompasses 25,000 km².

In regions further removed from the Atlantic coast, heath formation presents no danger since, despite the fact that *Calluna* has very small leaves with a thick cuticle and that the stomata lie in a hairy groove, heather spp. are very sensitive to frost. The leaves of *Calluna* differ from true xeromorphic leaves in that the mesophyll has a very loose structure. Transpiration is relatively active in summer if the water supply is adequate, and in shady habitats can equal that of *Oxalis acetosella*, calculated on a fresh weight basis. But when water is scarce, transpiration is sharply decreased. These properties, nevertheless, do not suffice to prevent water losses during the long periods of frost. Even in the mild winters in Heidelberg, in southern Germany, *Calluna* dries out because it lacks a protective covering of snow. In the north it is only found where there is a covering of snow each year.

Inland, in western Europe, heath is to be found in patches on the western slopes of the low mountains with an oceanic climate (Ardennes, High Venn, Eifel, Vosges, and even in the Black Forest on the Feldberg). It also extends as a narrow strip along the southern coast of the Baltic.

4 Deciduous Forests as Ecosystems or Biogeocenoses

A deciduous forest is a multilayered plant community, often consisting of one or two tree strata, a shrub stratum, and a herbaceous stratum. Numerous hemicryptophytes grow in the latter, as well as many geophytes which develop only in the spring. The poor illumination on the forest floor does not favor the development of therophytes, i. e., annuals. A mossy ground cover is lacking since it would be covered by the falling leaves, and therefore mosses are only found on rocks or tree stumps projecting above the ground.

There are no virgin forests in the European nemoral zone, and the structure of the forests is determined by the type of management practiced. From the forestry point of view it is the woody species that are of importance, and the herbaceous layer is only indirectly managed. If forest grazing is practiced, on the other hand, it is the herbaceuos layer which is changed by selective grazing of cattle which also are a danger to the tree saplings. High forests (Hochwald) run on an rational basis approach virgin forest, although they differ basically in the small number of species in the tree stratum, in thef act that the trees are all of the same age, in the lack of rotting wood on the forest floor, and in their homogeneity of structure. Virgin forest is usually of a mosaic-like structure.

Cultivated beech forests are pure stands with only a herbaceous stratum in addition to the trees. Oak forests, on the other hand, are usually mixed stands of various deciduous tree species and possess a shrub stratum. Of the various types of deciduous forest biogeocenoses, a western mixed forest in Belgium (p. 16) and eastern oak forests on the forest-steppe margin have been investigated in detail. Further investigations have been carried out, but the results are not yet available (projects of the International Biological Program). The active layer in deciduous forests is the tree canopy, where both direct radiation from the sun, as well as diffuse radiation, are to a large extent transformed into heat. Of the incoming radiation 17 percent is reflected by trees in leaf and only about 11 percent in the leafless condition, but in both cases less radiation is reflected than from meadows and cultivated land (25 percent). Only a minute portion of the daylight penetrates into the forest vegetation. The following figures apply to eastern European oak forests (13-to 220-year-old stands).

In young forests in full leaf only 1.2 percent of the daylight penetrates half way down and 0.6 percent to the forest floor, whereas in very old forests the figures are 20 percent and 2 percent, respectively. The mean daily temperature of the canopy in summer is 2° C higher than that on the forest floor, the mean daily maximum is 11° C higher, and the mean daily minimum 3° C lower. The mean humidity of the air is 98 percent on the ground and drops as low as 77 percent with increasing height. The wind velocity in the forest is low, and since the forest floor is protected from direct radiation, the air in the forest remains cooler during the day than in an open situation.

The crowns of the trees intercept 11—12 percent of the rain, while the remainder either drips down or runs down the trunks. Where snow has accumulated on the forest floor in winter it melts only in spring. The resulting water seeps almost entirely into the litter layer, whereas on open ground it would run off the surface of the still-frozen ground. In summer, transpiration of the tree stratum is so intense that the ground water receives no additional water from forested areas. Water losses from the herbaceous stratum are 5—6 time smaller. A well-developed deciduous forest of the forest-steppe region uses nearly all of the incoming precipitation whereas a beech forest in central Europe uses only 50—60 percent, although in summer months there is no surplus.

The productivity of a forest depends to a large extent upon its Leaf Area Index (LAI), that is to say, upon the ratio of total leaf area of the stand to the ground area covered by it. This ratio is limited to a maximum value, above which it may not rise, since otherwise the lower, overshadowed leaves would not be able to maintain a positive balance of assimilation. This maximum, however, not only depends upon light intensity but also decreases in the face of inadequate supplies of water or nutrients. The LAI of a pure oak stand is 5 (higher in wet years), and in mixed stands with a good water supply, inclusive of all tree species and shrubs, it can exceed 8.

The productivity of the tree stratum of a central-European beech forest has been studied in the greatest detail. The results are expressed in tons of dry substance per hectare and year for a 40-year-old population (from Müller and Nielsen (10)).

Gross production of the assimilating leaves $\quad\quad$ = 23.5 t/ha
Respiratory losses: leaves 4.6 t, stems 4.5 t, roots 0.9 t $\quad$ = 10.0 t/ha

Annual production of leaves (2.7 t), stems (1.0 t)
   litter and roots (0.2 t)   = 3.9 t/ha
Wood production, aboveground 8.0 t/ha and
   underground 1.6 t/ha   = 9.6 t/ha

On an average, 6 t/ha of the maximum of 8 t/ha of trunk wood is utilizable, which is equivalent to 11 m³. The same weight of wood is produced by the spruce, but it occupies a mean volume of 17 m³. Similar figures have been calculated for an east European deciduous forest with LAI = 4.5, a primary production of 12 t/ha, and a wood production of 5 t/ha, as well as for an oak-pine forest on Long Island in New York state. If the chemical energy bound in the process of production is related to the incoming radiation energy for one hectare of forest, the figures of 2 percent for gross production and 1 percent for the primary production are obtained. One-third of the incoming energy is used up in transpiration, and the remainder is transformed into heat. The leaf mass and leaf area formed annually increase rapidly during the first 20 years, but as soon as the canopy closes, leaf mass and LAI remain almost constant. The height of the canopy above the ground is all that alters with upward growth of the trunks. The litter consists of the old leaves and twigs, and these, together with dead roots, constitute the total litter fall of the forest.

Only the new wood is accumulated, so that the standing phytomass of the forest steadily increases even with great age and may exceed 200 t/ha for a 50-year-old stand and 400 t/ha for a 200-year-old stand.

The following figures for average yearly wood production related to age (in brackets) were found for eastern European oak forests

3.8 t/ha (13), 3.6 t/ha (22), 4.3 t/ha (42), 4.7 t/ha (56), 0.4 t/ha (135), 0.0 t/ha (220).

Litter, too, accumulates in the forest until a state of equilibrium is reached, at which time as much litter is mineralized annually as is produced in the course of one year. A portion of the most important minerals is bound in the litter (N, P, Ca, K), and therefore thick layers of raw humus are undesirable. Removal of litter by farmers for use in stables is particularly detrimental because the resulting loss of nutrients, particularly of calcium, leads to impoverishment and acidity of the forest soils and a consequent decrease in wood production. Decomposition of litter also involves the mineralization of nitrogen compounds. The larger part of the nutrients in the lower, decomposing humus layer is available to the tree roots which are therefore abundant in this soil horizon. Next to the water supplies the soil fauna is very important for forest productivity, whereas the part played by animal organisms above ground is negligible, and only a few percent of the biomass are accounted for by insect feeding (see p. 16).

5 The Ecophysiology of Deciduous Broad-Leaved Trees

The size of a tree renders it an unsuitable object for experimentation. Its form is to a large extent dependent upon its surroundings. The crown of a solitary tree is usually dome-shaped or spherical, whereas it is usually very small if the tree is part of a dense stand. Since the leaves are arranged in several layers, the outer ones are exposed to the full daylight while the inner ones grow in the shade. A distinction is therefore made between sun leaves and shade leaves, which differ in anatomical-morphological and ecophysiological properties and are linked by intermediate forms. Sun leaves are smaller and thicker, have a denser nervature, and possess more stomata on the under surface per mm². In other words, they are more xeromorphic than the large thin shade leaves.

The structural differences result from the unfavorable water balance at the time when the buds for the next spring are being formed. The twigs exposed to the sun transpire more actively, as is registered by an increase in cell-sap concentration. The cell-sap concentration of the sun leaves of a beech tree amounts to 16.3 atm and of the shade leaves 11.6 atm. Their CO_2 assimilation also differs. In laboratory experiments it has been established that in darkness the shade leaves respire less actively per dm² surface area than the sun leaves. In a beech, the shade leaves give off only 0.2 mg CO_2 per dm²/hr as compared with 1.0 mg for the sun leaves. This explains the finding that in spring the light compensation point (where respiration = gross photosynthesis) of the shade leaves is 350 lux and of the sun leaves 1,000 lux. Photosynthesis increases proportionally to light intensity until a maximum is reached, which for the shade leaves is 20 per cent of the maximum daylight and for the sun leaves about 40 percent. Thus, the shade leaves are better able to utilize the lower light intensities and the sun leaves the higher intensities, although even the sun leaves do not appear to utilize the available daylight to the full. The above figures apply to leaves oriented at right angles to the incoming light, whereas the sun leaves near the apex of the tree are nearly always rather steeply inclined. This protects them from overheating by the sun and thereby helps to reduce water losses. It also means that more light can penetrate the outer canopy to the advantage of the lower leaves. The leaves in the deep shade are always at right angles to the incoming light, and even with a LAI of 5 or more a mean positive production is made possible.

If illumination is continuously below a certain minimum, respiration is no longer compensated for by photosynthesis, losses of material occur, and the leaves turn yellow and are shed. This minimum, expressed in percent of the full daylight, varies from one tree species to the next. A "shade" tree with a dense crown, such as beech, has a low light minimum (1.2 percent) and "light" trees, such as birch and aspen, with a thinner crown have a higher light minimum (11 percent). The figures for species such as maple and oak lie somewhere between those mentioned above. This light minimum is valid for the crown of the tree and does not necessarily coincide with that light minimum to be exceeded if the tree seedlings are to develop on the forest floor, although the values run parallel with one another. Beech seedlings require little light, whereas birch seedlings need at least 12—15 percent of the total daylight.

Light conditions are of vital importance to trees in competition with one another. Light-demanding trees can mature within a few years in a clearing, and under their canopy shade-tolerating trees germinate and gradually grow higher, producing in turn a canopy so dense that the "light" trees cannot produce any reserves and are incapable of regeneration. In time, it is the species tolerating the most shade that achieves dominance, other habitat factors being suitable.

The zonal forests in central Europe consist of beech *(Fagus sylvatica)*. Only on very poor soils, or where the ground water is high, or in the driest type of valley is it unable to compete successfully. In the western parts of eastern Europe the climate is too continental for the beech, and it is replaced by another shade-loving species, the hornbeam *(Carpinus betulus)*, and still further east by the oak *(Quercus robur)*.

6 Ecophysiology of the Herbaceous Layer

The microclimate of the forest floor is vastly different from that of an open habitat. When the forest is in leaf, the light intensity on the forest floor is weaker, the temperature more moderate, and the humidity of the air and upper soil layers greater than that outside the forest. For these reasons the herbaceous plants of the forest are all shade-tolerating and hygrophytic, and their cell-sap concentration is very low, which means that the hydrature of their protoplasm is favorable.

On a clear day, light conditions on the forest floor can be very heterogeneous. Single rays of sunshine falling through the tree canopy cause sun flecks, and as the sun moves across the sky or the branches are moved to and fro by the wind these flecks continuously change their position and intensity.

If a leaf of a herbaceous plant is hit by a sun fleck, its illumination can rise by 10- or even 30-fold, a factor of great significance for the plant's photosynthesis. It is therefore preferable in determining the amount of light received by herbaceous plants in percent of the total daylight to carry out the comparative measurements on a bright day with more or less even cloud cover. Such measurements provide us with no more, however, than preliminary information. It would be better to measure the sum total of the daylight falling on a certain place on the forest floor with the aid of a self-registering light meter.

Before the trees come into leaf the herbaceous stratum is very adequately illuminated, but as the trees come into full leaf the situation gradually deteriorates. The following values were obtained between March and June in an oak-hornbeam forest:

On 12 March—52%, on 15 April—32%, on 10 May—6.4%, on 4 June—3.7%.

For a beech forest the values were as follows:

Mid-May—6%, a week later—3%, on 7 June—1.5%.

The favorable light conditions prevailing before the trees come into leaf are exploited by the spring geophytes (*Galanthus, Leucojum, Scilla, Ficaria, Corydalis, Anemone*, etc.). They profit from the fact that even in April the litter layer in which they root is warmed up to 25—30° C because of its uninterrupted exposure to the sun's rays. Geophytes benefit from the fact that heat capacity of the air-containing litter layer is small, and, as a result, its temperature (not heat!)

conductivity is very good. The trees come into leaf later because the deeper soil layers in which they root are slow to warm up. In the short, early spring season geophytes flower and fruit and store up reserves in their underground storage organs for the coming year. When the trees are in leaf, the leaves of the geophytes turn yellow and a dormant period begins for them. The death of the leaves is not, however, due to the deeper shade, but is the expression of an endogenous rhythm; they die even sooner in light. Apparently they are just the right plants to fill the vacant ecological niche in deciduous forests.

The majority of species of the herbaceuous layer are hemicryptophytes, which means that their regenerative buds form at the base of the shoot and spend the winter immediately beneath the ground surface, protected by a covering of autumn leaves and sometimes of snow as well. Since they begin to grow after the trees have come into leaf, they have to live under unfavorable conditions. They avoid the deepest forest shade, and therefore it is possible to determine illumination maxima (L_{max}) and minima (L_{min}) for such plants. The following limiting values in percent of total daylight are given as examples (L_{max}—L_{min}):

Lamium maculatum 67—12%, *Lathyrus vernus* 33—20%, *Geranium robertianum* 74—4%, *Prenanthes purpurea* 10—5% (sterile to 3%).

L_{max} is dependent upon the water supply; hygrophilic species require damp soil and cannot tolerate a high saturation deficit of the air such as would occur under conditions of full illumination, although, in mountainous regions, where the humidity is higher, they are often found growing in the meadows rather than in the forest. The fact that the species of the forest floor cannot compete with sun-loving plants in open habitats may also be responsible for L_{max}.

A few species otherwise capable of tolerating full daylight can also be found in the forest, although they are there poorly developed and often sterile (the strawberry for example).

If the trees are felled, such plants develop luxuriantly in the clearings, flower, and bear abundant fruit. L_{min} represents the starvation limit, the light just sufficing for the production of the organic material indispensable for development. Plants which remain sterile manage with less light, as do the cryptogams (ferns and mosses). In general, the so-called "dead" shade of the forest commences at an illumination of 1 percent, below which point only the fruiting bodies of heterotrophic fungi are to be found.

As the edge of the forest is approached and light conditions improve, the plants are better developed. Productivity of the herbaceous layer increases linearly with the mean light intensity on the forest floor.

The illumination in percent of daylight is not a good means of estimating photosynthesis because in the presence of heavy cloud the light may be very weak, whereas on a clear day it can be very much stronger. This is the reason why the production of the plants of the forest floor is negative in dull weather or, in other words, the loss of carbohydrates by respiration in 24 hours is higher than the production by photosynthesis during such a day. On bright days the balance is positive. It is only essential, however, for production to be positive over the entire season of growth. Certain plants of the forest floor (tree seedlings, *Oxalis, Asperula, Asarum, Viola,* and ferns) adapt themselves to the unfavorable light conditions prevailing after the trees are in leaf by drastically reducing their respiration when they reach the starvation point. This results in a lowering of the light compensation point and better production.

In other species with evergreen leaves the point of light compensation remains unaltered. They live from their reserves in the summer and replenish them in autumn or spring, when the forest is bare *(Stellaria holostea, Hedera,* etc.).

Another factor of importance in the herbaceous layer is competition from the roots of the trees. Water plays a vital role in the dry forest regions bordering the forest-steppes. Trees, with their higher cell-sap concentration, are able to develop larger suction tensions in their absorbing roots, so that they are better able to obtain water from the soil than are the herbaceous plants. As a result, the floor of beech forests is bare *(Fagetum nudum).* If the roots of the trees are severed, thus excluding them from competition, herbaceous plants develop, proving that water and not light was the limiting factor.

On very shallow soils trees also extract the nutrients, especially nitrogen. Herbaceous plants are obliged to make do with what is left. Only plants with low nutrient requirements, like *Luzula luzuloides, Deschampsia flexuosa, Potentilla sterilis, Vaccinium myrtillus, Calluna vulgaris,* etc., are found in such forests.

7 The Effect of the Cold Winter Period on Plants of the Nemoral Zone

Plant damage occurring in a cold winter can be due to one of two causes:

1. Direct damage due to freezing of tissue water. This is termed frost damage.

2. Drying-out of aerial organs, which even at low temperatures transpire to a certain degree. When the conducting vessels are blocked by ice, insufficient water reaches the aerial organs to meet these transpiration losses, and the organs dry out. This is called frost drought.

Plants are not equipped with any means of protection against the effects of low temperature, their own temperature always being that of the surrounding air. Their only possible adaptation is to become hardened. If the resistance of plant organs to cold is tested in summer by placing them in a refrigerator at various temperatures below 0° C for 2 hours, it can be shown that only a few degrees suffice to cause irreversible damage. The same plant organs tested in winter, however, tolerate much lower temperatures without undergoing damage because they have developed a so-called *hardiness*. This is a physiological process taking place in autumn with the onset of the first cool nights. In spring when the weather becomes warmer, the opposite process is initiated (dehardening).

Hardening is connected with certain physico-chemical changes in the protoplasm which are as yet not fully understood. It is accompanied by a sudden increase in cell-sap concentration of several atmospheres because of an increase in sugar concentration. Protoplasm in its hardened state is more or less inactive. The resistance to cold of the buds of European deciduous trees in their winter condition can increase from − 5° C in autumn to − 25° C or even − 30° C in January or February. A larger increase in resistance to cold is developed in a cold winter than in a mild one. Among related species of a genus the resistance is the greater the further the species advances into the continental region.

As a rule "hardiness" suffices to prevent frost damage to the native trees in Europe in a cold winter, although exotic species imported from warmer climes often suffer. If an early frost occurs before hardening has set in, or if there is a late frost after dehardening has commenced, frost damage is widespread. Late frost damage is most commonly found in young, just opened leaves, and it can also cause cambium damage if the trees are already running sap and the protoplasm in an active state.

The eastern limit of the distributional area of the beech may possible be conditioned by the frequent late frost damage which reduces the powers of competition of the trees. An increase in resistance to cold by hardening can also be demonstrated in herbaceous forest plants, even though they are not exposed to extremely low temperatures because of their covering of litter and snow. Resistance to cold in the evergreen leaves of, for example, *Anemone hepatica* reaches only to $-15°$ C, that of the better protected flower buds to $-10°$ C, and of the rhizomes only to $-7.5°$ C.

Damage due to frost drought is more difficult to detect. Loss of the intensely respiring leaves, bud protection by hard bud scales, and protection of the twigs by a layer of cork prevent the loss of large quantities of water by deciduous trees in winter. Nevertheless, a certain amount of transpiration can be measured in the bare twigs in winter. It is higher in deciduous trees than in evergreen conifers, and higher in the deciduous species in the south than in those of northerly distribution. These transpiration losses become dangerous in spring, when the intensity of the incoming radiation increases and the air temperature rises while the ground is still frozen hard. Buds and twigs sometimes dry out as a result. Evergreen broadleaved trees and the broom-like shrubs such as *Sarothamnus* or *Ulex* are especially sensitive in this respect.

Frost damage usually occurs at the coldest time of the year, but damage due to frost drought is more common as spring approaches, and on warm southern slopes. It should not be confused with late frost damage.

8 The Various Deciduous Forest Regions

All examples so far quoted have been taken from European deciduous forests. Since they contain very few species, their tree stratum often, in fact, consists of a single species, they are simpler to study than others. The beech forests of the central European lowlands are found in the montane belts of the Pyrenees, the Alps, the

Bosnian mountains, and in the Mediterranean regions wherever there is a definite cloud belt, as is the case, for example, in the uppermost forest belt of the Appenines between 1,000 m and 1,700—2,000 m. The timberline here is formed by beech. The climatic conditions in these altitudinal belts correspond, at least as far as temperature is concerned, to the lower belts in central Europe, and since in summer this belt is also covered by cloud, the hydrature conditions, too, are similar. Whereas the limits of the various altitudinal belts are set by the competitive qualities of the plants as a result of the prevailing climatic conditions, competition plays no part at the tree limit. Annual production decreases steadily with altitude owing to the shortened period of plant activity, and thus the maturation of viable seeds is reduced. The ultimate decisive factor at the timber line formed by beech seems to be the damage caused by late frosts. Almost every year it kills off already opened buds and further weakens the stunted trees. Nowadays, however, the timber line is largely influenced by anthropogenic factors, such as grazing and damage done to the young foliage by animals.

The conditions described for Europe hold equally for the deciduous forests in eastern North America and Asia. The numerous species in the tree stratum of these forests, however, render ecophysiological investigations extremely difficult, and comprehensive publications are still lacking. The complicated floristic composition of such forests cannot be dealt with here.

VII The Arid Vegetational Regions of the Temperate Climatic Zone

1 Forest-Steppe as a Semiarid Transitional Zone

The deciduous forests of the temperate zone are confined to climatic regions of an oceanic nature, where there are no sharp extremes of temperature and the rainfall is more or less evenly distributed throughout the year, usually with a summer maximum. Steppe and desert occupy the continental regions, which are much more extensive in the Northern Hemisphere. In a continental climate the temperature amplitude is greater and the summers are hotter, but the winters

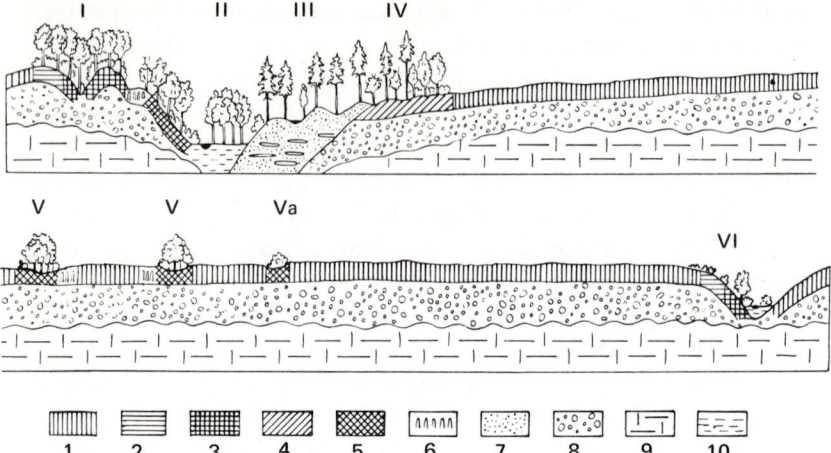

Fig. 61. Relation between vegetation, soil and relief in the forest-steppe (greatly modified, from Tanfiliev and Morosov). 1. Deep, poorly-drained chernozem with meadow-steppe, 2 degraded chernozem and 3 dark grey forest soil (both well drained), 4 porous sandy-loamy forest soil, 5 light-grey forest soil, 6 solonetz on flat terraces or around depressions with no outflow and with soda accumulation, 7 fluvioglacial sands, 8 moraine deposits or loess-like loam, 9 pre-glacial strata, 10 alluvium in the river valleys. I Oak forest on well-drained elevations or on slopes; II flood-plain forests (oak, etc.); III pine forests on poor sands with *Sphagnum* bog in wet hollows; IV pine-oak forests on loamy soils; V Aspen groves in small hollows (pods), in spring containing water that slowly seeps away (soil in central portion is leached); V a the same, but with willows; VI ravine-oak forest, with steppe-shrubs at upper margins.

much colder so that the annual mean temperature is lower than in oceanic regions. This is accompanied by a decrease in annual rainfall and more arid summers.

The forest-steppe zone represents a transition from deciduous forest to grassy steppe. It is not a homogeneous vegetational formation like the climatically conditioned tropical savanna, but rather a macromosaic of deciduous forest stands and meadow-steppe. At first the former predominate, the steppes forming scattered islands. But the more arid the climate becomes, the more does the situation tend to be reversed, until finally small islands of forest are left in a sea of steppe. The climate itself is no more favorable to forest than to steppe, so that relief and soil type (Fig. 61) determine the predominating vegetation. Forests are found on well-drained habitats, on slightly raised ground, on the sides of the river valleys, and on porous soils, while the meadow-steppes occupy badly drained, flat sites with a relatively heavy soil. The grasses and the tree seedlings compete with one another and if, as happens in the course of re-forestation experiments, the young tree plants are protected for the first couple of years from competition with the roots of the grasses, they are able to grow on the steppe, although they are incapable of regeneration. In previous times fires caused by lightning and the grazing of big-game herds encouraged the growth of the steppe, which nowadays, however, is entirely given over to arable farming.

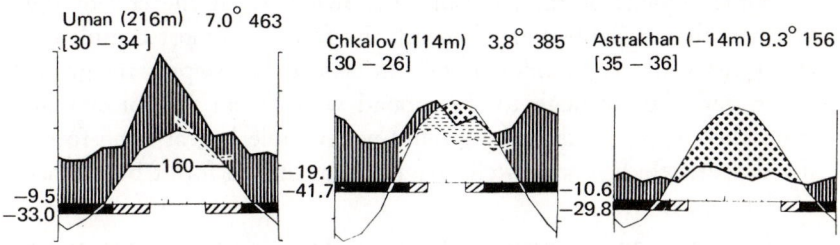

Fig. 62. Climatic diagrams from the forest-steppe zone (with dry season), from the steppe zone (with drought and longer dry season) and from semi-desert (with long summer drought). (See also Fig. 7, Odessa.)

The forest zone, forest-steppe zone, and steppe zone of eastern Europe are readily distinguishable from one another on a climatic basis. Climatic diagrams for the forest zone reveal the absence of a period of drought, whereas diagrams for the steppe zone always indicate the presence of such a period. Although no drought is to be detected on the diagrams for the forest-steppe zone, a dry period is always recognizable, which is not the case for the forest zone (Fig. 62).

During the postglacial period the boundary between forest and steppe shifted. In the soil profile beneath the present-day forest stands, it is possible to find "Krotovinas", the deserted burrows of steppe rodents (suslik, or European ground squirrel, *Citellus citellus).* Since these animals never inhabit forests, it must be assumed that before the forest-steppe was inhabited by man the forests were in the process of advancing, probably on account of the climate's becoming more humid. Later shifts in the boundaries cannot be detected because of the large degree of human interference.

The replacement of the forest zone by steppe in continental regions is governed by the supply of water.

It has already been mentioned that deciduous forests in the vicinity of the "dry" limit are lacking in undergrowth since the trees appropriate all of the available water. Any further decrease in rainfall or increase in summer transpiration values means that the water supplies can no longer support a forest vegetation although they suffice for grass, which can make do with less. In forest-steppe, where the two types of vegetation grow side by side and therefore receive the same amount of rain, it can be shown that, at the end of the summer, the available water beneath the forest is almost completely exhausted while small amounts remain beneath the steppe. In spring, less water seeps through to the ground water from the forest than from the steppe, so that the ground-water table beneath the forest lies deeper, which is why wells are not sunk in the forest but rather on the steppe.

In August and September the grass steppe dries out, the water supplies being insufficient to cover losses due to transpiration. This is not harmful to the grasses, although the trees suffer damage if they lose their leaves too early or if entire branches die off.

The amount of water required by a forest increases with age. Experiments in reforestation have revealed that young forest plantations grow relatively well, but that with time the tips of the older shoots dry off and fresh shoots are then put out from below. The trees therefore develop abnormally as a result of the water shortage. If, however, ground water is also available, healthy stands develop. Savanna-like communities are missing in the forest-steppe because the individual deciduous species are unable to compete successfully with the grasses. Only low shrubs such as *Spiraea, Caragana,* and *Amygdalus* are more common, although they are generally to be met with on stony ground less suited to the dense root systems of the steppe grass (p. 68).

2 Soils of the East European Steppe Zone

The east European steppes are the cradle of the science of soil types, the foundations of which were laid by Dokutchaev (1883) and Glinka (1914). There is no other region of comparable area where the parallel zonation of climate, soil type, and vegetation can be seen so clearly. It must be added, however, that very little remains of the original vegetation. The conditions responsible for the clear zonation are the extreme uniformity of relief and the fact that the parent rock is to a large extent homogeneous (loess). The climate alters steadily from northwest to southeast, the summer temperatures

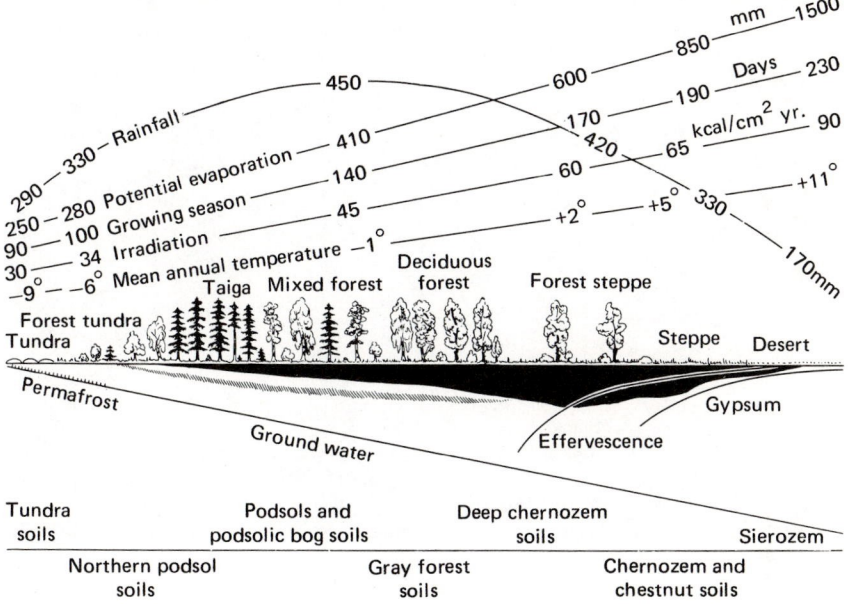

Fig. 63. Schematic climate-, vegetation- and soil profile of the east-European lowlands from N. W. to S. E. (from Schennikov, modified). Black = humus horizon, diagonal shading = illuvial B-horizon.

and potential evaporation rising, while the rainfall decreases, so that aridity becomes more and more pronounced. The boundary between forest zone and forest-steppe zone coincides with the boundary between humid and arid regions. This means that to the north of this demarcation line the annual rainfall exceeds the potential evaporation, whereas to its south the latter is the higher of the two (Fig. 63). In depressions with no outflow saline soils therefore form.

The distribution of the various soil types is depicted in a much simplified form in Fig. 64. Humid regions have a typical podsol soil

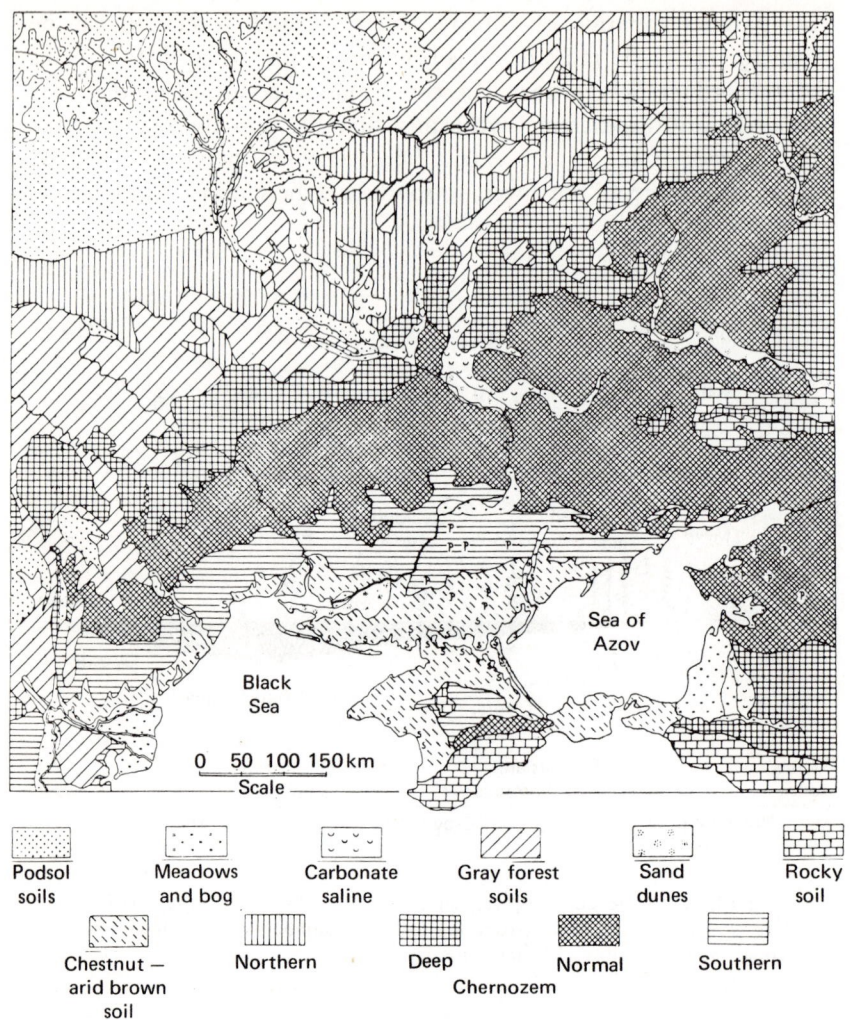

Fig. 64. Soil map of the east-European steppe region and the adjacent forest regions. P = Pod (hollows with no outflow, in the steppe), S = saline soils (solonchak).

and slightly podsolized gray forest soil, whereas in the arid regions the soil ranges from chernozem to chestnut and arid brown (burozem). The soil types are recognizable from their soil profiles, which are shown in Fig. 65.

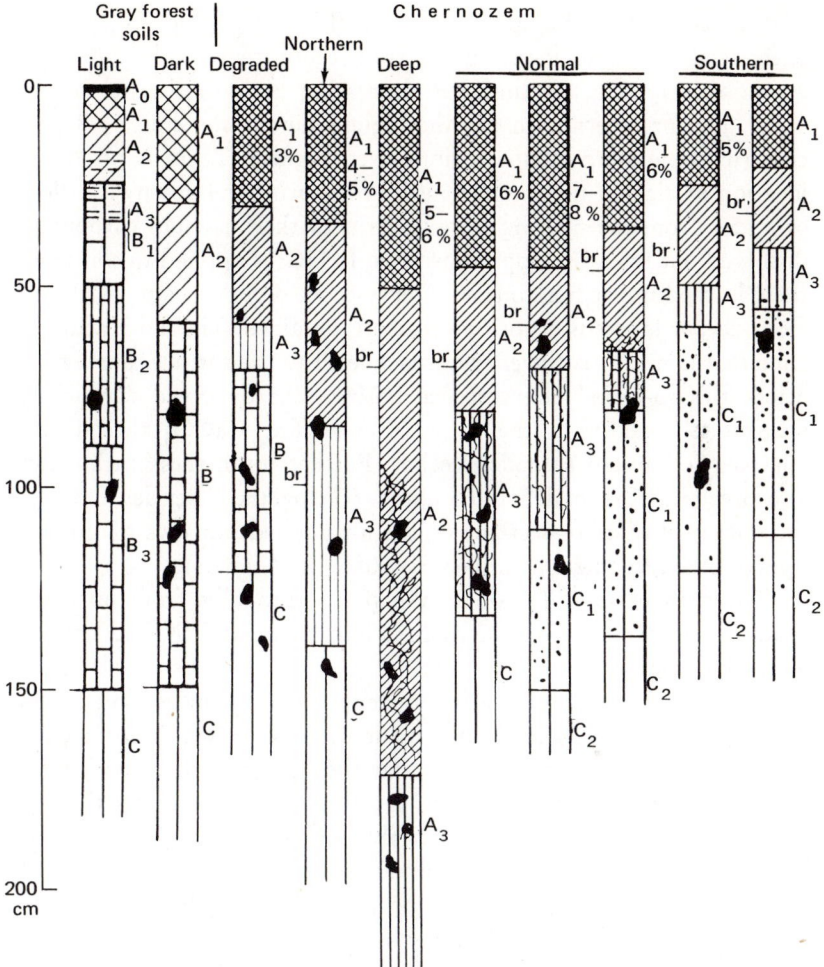

Fig. 65. Schematic representation of the soil profiles of the forest-steppe- and steppe zones (west of the Dnieper) from N to S. Percentage figures = humus content of the A_1 horizon, br = effervescence horizon, wavy lines = pseudomycelia ($CaCO_3$), small dots = $CaCO_3$ nodules, large black spots = krotovinas (abandoned groundsquirrel burrows), horizontal dashes = laminated structure in forest soil.

The chernozems are A—C soils, or pedocals, without a clayey illuvial horizon (B). The zones are subdivided as follows: northern-, thick-, normal-, and southern-chernozems. The humic A horizon consists of a black A_1 layer, a slightly lighter A_2, and a loess layer

167

slightly colored by humus A_3. Below this is C, which is original, unchanged prismatic loess. In the thick chernozem the humus layer goes down 170 cm, its thickness decreasing both to the north and south. Normal chernozem has the highest humus content, 7—8 percent (in the eastern steppe regions even higher). There is no translocation of clay in the chernozems, but in spring the downward flow of water from the melting snows carries with it calcium carbonates dissolved out from the upper horizons. If HCl is applied to soil from these leached upper horizons no effervescence occurs — only with soil from deeper layers is a positive reaction obtained. The more arid the climate, the nearer is the effervescence level to the surface. Somewhat below this so-called effervescence horizon the dissolved carbonates precipitate, usually in the form of vary fine $CaCO_3$ threads reminiscent of mould (pseudomycelia). Further south these carbonates also precipitate as small white nodules (Bjeloglaski). Besides $CaCO_3$, the humus-filled cross-sections of the abandoned burrows of ground squirrels are recognizable in the soil profile (krotovinas).

The changes in soil profile take place gradually from north to south in conformity with the changes in climate, and they reflect the increasing aridity.

Beneath the forest of the forest-steppe zone the upper soil layers remain wetter. The A_0 horizon is made up of litter which mixes only slightly with the mineral soil, so that the humus horizon under the moist hornbeam forests *(Carpinus)* is a light gray, and under the dry oak forests dark gray. The good friable structure is lost, and the soil becomes laminated. Beneath the humus layer there are mealy, bleached sand grains and below this a compact B horizon, indicating the beginning of podsolization. There is hardly a trace of this, however, in the degraded chernozem beneath the shrubby oaks which constitute the last outposts of the forest. Below the moist humid parts of the meadow-steppes, the soil is typical northern chernozem with a very deep effervescence level and no $CaCO_3$ precipitations.

On the basis of the surviving remnants of natural vegetation it has been possible to show that every soil type has its corresponding plant community, as in the following summary:

Soil type	Vegetational unit
Gray forest soil	Oak-Hornbeam- and Oak-forest
Degraded chernozem	Oak-Blackthorn bush *(Prunus spinosa)*

Northern chernozem	Damp meadow-steppes with abundant herbs
Thick chernozem	Typical meadow-steppe
Normal chernozem	Feather-grass *(Stipa)* steppe with abundant herbs
Southern chernozem	Dry *Stipa*-steppe, few herbs

This scheme makes it possible, with the aid of a soil map, to reconstruct the original vegetation.

3 Meadow-Steppes on Thick Chernozem and the Feather-Grass Steppes

The word "steppe" comes from the Russian word "step", and therefore its use should be confined to those grass steppes of the temperate zone that are comparable with the east European steppes. No steppes of this type occur in the tropics (p. 67), where it is more appropriate to refer to tropical grassland. The word steppe often conjures up a picture of dreary, poor vegetation, although the very opposite holds true for the northern variant of the east European steppe. Nowadays these are the most fertile parts of Europe, with the best chernozem soils. In their natural condition they excel even the lushest European meadows in the abundance of their colorful blossoms. Only in autumn do they give the impression of dryness.

It has already been said that the forest-steppe is a macromosaic of deciduous forests and meadow-steppe. Since the deciduous forests have already been dealt with (p. 175), the meadow-steppes will now be considered in more detail.

The seasonal course of events is as follows. When the snow melts, the steppe soil is thoroughly wet, the temperature rises, and a profusion of spring flowers develops. At the end of April the mauve blossoms of *Pulsatilla patens* appear, *Carex humilis* begins to shed its pollen, and at the beginning of May they are joined by the large golden stars of *Adonis vernalis* and the pale blue inflorescences of *Hyacinthus leucophaeus*. By mid-May the steppe is verdant, and *Lathyrus pannonicus*, *Iris aphylla*, and *Anemone sylvestris* are in flower among the sprouting grasses. The most colorful stage is reached at the beginning of June, when innumerable *Myosotis sylvatica*, *Senecio campestris*, and *Ranunculus polyanthemus* are in bloom. At this point the first plumes of *Stipa joannis* appear, and by early summer the long feathery awns of the various *Stipa* species, interspersed with the panicles of *Bromus riparius* (closely related to *B. erectus*), are swaying in wave-like motion in the wind. Intermingled are the blossoms of *Salvia pratensis* and *Tragopogon pratensis*. Toward the end of June the flowers of *Trifolium montanum, Chrysanthemum leucanthemum,* and *Filipendula hexapetala* whiten the steppe, a colorful contrast being provided by *Campanula sibirica, C. persicifolia, Knautia arvensis,* and *Echium rubrum.* At the beginning of July, when *Onobrychis arenaria* and *Galium verum* come into flower, the glorious colors begin to near their end.

From mid-July onward the plants begin to wither despite the appearance of the dark blue panicles of *Delphinium litwinowi* and, later, the red-brown candles of *Veratrum nigrum.* From August onward the steppes look dry and remain in this state until they are covered with snow.

This description shows that the dry meadows and steppe-heath of Central Europe merely represent meager extrazonal outposts, on dry shallow habitats, of the meadow-steppes of humid climates. In floristic composition the two are very similar, except that in central Europe sub-Mediterranean elements, such as orchids which are not found on the steppes are also present.

Further to the south of the meadow-steppes of the forest-steppe zone is the feather-grass steppe, on normal and southern chernozems. Various species of *Stipa* predominate, and, faced with increasing dryness, the less drought-resistant herbs are incapable of successful competition and gradually recede. The density of the plant cover decreases to such an extent that the ground is covered in spring with the moss *Tortula (Syntrichia) ruralis* and the alga *Nostoc.* In spring geophytes such as *Iris, Gagea,* and *Tulipa,* and some winter annuals *(Draba verna, Holosteum umbellatum)* are more abundant. *Paeonia tenuifolia* is especially striking. Other herbs make their appearance in summer *(Salvia nutans, S. nemorosa, Serratula, Jurinea, Phlomis,* etc.) and are joined later by Umbelliferae *(Peucedanum, Ferula, Seseli, Falcaria)* and Compositae *(Linosyris).*

Further to the south the density of the vegetation decreases still more. Apart from the feather-grasses, *Stipa capillata* and *Festuca sulcata* are more common and herbaceous plants with very long tap roots are frequent *(Eryngium campestre, Phlomis pungens, Centaurea, Limonium, Onosma).*

On the chestnut soils sagebrush *(Artemisia)* species become more abundant and initiate the transition to sagebrush semidesert.

4 North American Prairie

Although the conditions prevailing on the prairies and on the steppes are very similar, the situation in the former is more complicated. Whereas the steppes stretch at a latitude of about 50° from the outposts of the Carpathians far beyond the borders of Europe to the east, the prairies, although they, too, begin south of a latitude of 55° in Canada, extend in a north-southerly direction beyond a latitude of 30° and are succeeded by *Prosopis* savanna. Furthermore, the extensive plains of North America rise gradually to 1,500 m above sea level and the precipitation decreases from east to west, but the temperature rises from north to south. This means that there is no clear-cut soil zonation, but rather a checker-board arrangement of soil types (Fig. 66).

The individual vegetational zones, such as tall grass prairie, mixed prairie, and short grass prairie, succeed one another from east to west with increasing aridity, but within each zone there is a floristic gradient from north to south. *Andropogen* species, that is to say, grasses of southern origin, are more common in the prairie than *Stipa*.

In North America, too, there is a transitional zone of forest-steppe, in which the sides of the valleys and light soils are forested and the flat watersheds with heavy soils support grassland.

Tall grass prairie corresponds to the northern meadow-steppe on thick chernozem, but the prairie soils are wetter, the chalk is completely leached, and there is no effervescence level. The question as to why, nevertheless, no trees grow on the prairie has been settled experimentally by planting tree seedlings with and without the competition of grass roots. The results showed that if such competion is excluded, trees are, in fact, able to grow. Wherever prairie fires no longer occur, and where human interference is excluded, the forest slowly encroaches upon the prairie, with a bush zone in the vanguard, at a rate of about 1 m per 3 to 5 years. Statistics show that for 1965 an average of one fire caused by lightning occurred per 5,000 hectares of prairie, and it is clear that such fires are a natural environmental factor favoring the grasses. It must also be borne in mind that in earlier times the prairie vegetation was much favored by the large herds of grazing bison. A natural experiment was provided by the catastrophic drought of 1934—1941, the effects of which on the prairie vegetation were still evident in 1953. Such recurrent periods of drought every century are undoubtedly partially responsible for the absence of trees on the prairie.

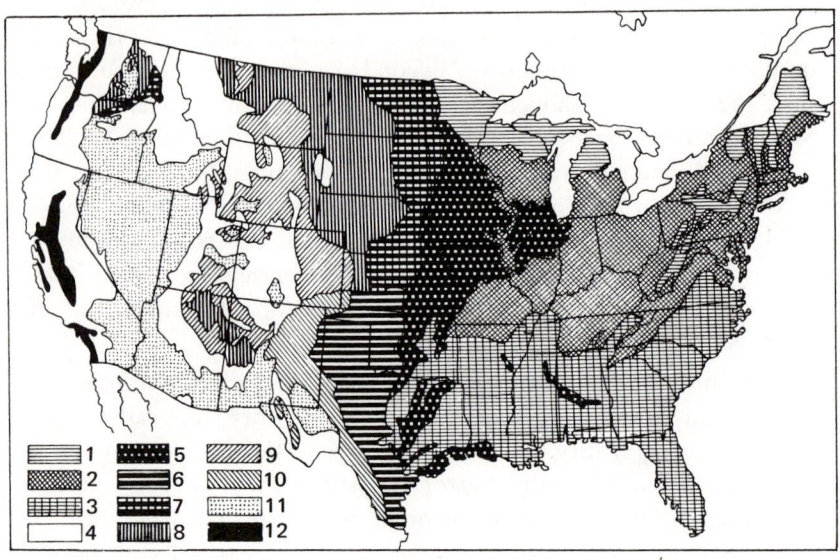

Fig. 66. Soil map of the U.S.A. (based on a map of the U.S. Dep. of Agric.) 1 Podsol soils, 2 grey-brown forest soils, 3 yellow and red forest soils, 4 mountain soils (general), 5 prairie soils, 6 southern chernozems, 7 northern chernozem, 8 chestnut-arid brown soils, 9 northern arid brown soils, 10 southern arid brown soils, 11 grey desert soils (sierozems), 12 Pacific valley soils. These correspond to: 1 coniferous forest zone, 2 and 3 mixed forests and decidous forest zone, 5 tall-grass prairie, 6—10 mixed- and short-grass prairie, 11 in northern part, sagebrush semi-desert, in southern part, other types.

Tall grass prairie is just as abundant in herbs as are the meadow-steppes and is floristically even richer. At the height of the flowering season in June, 70 species bloom simultaneously. The majority of the grasses *(Andropogon scoparius* and *A. gerardi)*, being southern elements, do not flower until late summer, and in normal years they are not troubled by lack of water since the prairie soils are moist to a great depth. The grasses themselves are 40—100 cm tall, even 1—2 m with their inflorescenses. In the mixed prairie zone, apart from the tall grasses *(Andropogon scoparius, Stipa comata)* there are many short grasses *(Bouteloua gracilis, Buchloe dactyloides)* which then assume the dominant role in the short grass prairie; herbaceous plants disappear, but *Opuntia polycantha* is abundant, particularly in overgrazed areas. Grazing tends to alter the appearance of the prairie slightly in the direction of an apparently greater degree of aridity, the tall grass prairie turning into mixed prairie, and this, later, into short grass prairie. The $CaCO_3$ deposits in the soil indicate an increasing aridity toward the west. In the short grass prairie $CaCO_3$ nodules are found at a depth of only 25 cm, the humus horizon is very shallow, and the plant roots are shorter since they only have to penetrate a short distance into the horizon with the chalk deposits, these being an indication of the mean depth to which the soil contains moisture.

5 Ecophysiology of the Steppe- and Prairie-Species

The cold winter on the one hand and the drought of late summer on the other limit the vegetational season of the steppe plants. Only about four months of favorable growth conditions are at their disposal, in spring and early summer. Most of the species are hemicryptophytes, and during this brief period they are obliged to build up a large productive leaf area at the smallest posible cost of material. Exact determinations of leaf area indices have not been made, but on the meadow-steppes the values are probably similar to those for deciduous forests. Nevertheless, the total leaf area varies greatly from year to year according to the rainfall. The figures for the aboveground phytomass of the feather-grass steppe, which is poor in herbs, are 4,530—6,250 kg/ha in wet years, as compared with 710—2700 kg/ha in dry years. This means that an insufficient supply of water is countered by a reduction in transpiring surface, and a consequently smaller productivity. The underground phytomass remains unchanged and is much larger than that aboveground:

Meadow-steppe: Phytomass 23,7 t/ha
                  (underground 84 percent)
                  Annual production 10.4 t/ha

Feather-grass steppe: Phytomass 20,0 t/ha
(undground 91 percent)
Annual production 8.7 t/ha

The aerial parts dying off each year form a litter layer on the ground (steppe felt), amounting to 8—10 t/ha on the meadow-steppe as compared with only 3 t/ha on the dry steppe. When the underground parts die, they are converted into humus by the soil organisms. In spring and summer the litter layer undergoes intense decomposition, with a minimum at the commencement of the drought period and a maximum at the beginning of winter. In protected areas, for example, the accumulation of too much litter is detrimental to the regeneration of the grasses, the plant cover becomes patchy, and weeds such as *Artemisia, Centaurea,* and thistles establish. If the steppe vegetation is to be maintained in its original form, a certain amount of grazing is therefore indispensable. This was provided in earlier times by gazelles and Saiga antelope, wild horses and donkeys, and, above all, by the innumerable steppe rodents (ground squirrels, etc.) and locusts. Earthworms and burrowing rodents contribute substantially to the mixing of the humus with the mineral soil. Occasional naturally occurring steppe fires led to the destruction of the accumulated litter. Nowadays, in the steppe reservations, the grass is mowed every three years for hay in order to reduce such accumulation.

A similar state of ecological equilibrium exists between the grasses and herbs in the steppe as between woody plants and grasses in the savanna (p. 68). Grasses possess a very intensive, finely branched root system, whereas that of the herbs is extensive, often with very long tap roots. The nature of the water economy of the steppe herbs places them among the group of malakophyllous xerophytes. Their cell-sap concentration is very low in spring. Temporary periods of dryness lead to wilting, accompanied by a sharp rise in cell-sap concentration. Late flowering species that bloom when the drought commences cut down their transpiration losses by allowing their leaves to wither. The flowers and fruits require but little water and receive the necessary building materials from the withering plant organs.

Typical of the open spaces of the steppes are the so-called Tumbleweeds *(Eryngium, Falcaria, Seseli, Phlomis, Centaurea, etc.).* The rigid stem supporting the spherical dried-out inflorescence breaks off at the root collar and is rolled across the steppe in the wind, scattering its seeds as it goes. The rolling plants often get entangled with one another and form enormous masses which are driven in leaps and bounds at great speeds across the steppes.

Stipa species regulate their transpiration by rolling up their leaves as well as closing their stomata, in this way also reducing photosynthesis. The distribution of the various species is determined by their adaptation to specific habitat conditions.

The water economy of the steppe-heaths of Central Europe has been the subject of many investigations. This type of vegetation, an extrazonal relict of a xerothermic period of the postglacial era, is confined to warm loess- or calcareous-slopes or to sandy soils, and harbors very hydrolabile malakophyllous steppe species. The dryness of such biotopes in Central Europe is due rather to the small field capacity of the soils and to the high potential evaporation on southern slopes than to the climate itself. Instead of a long drought in the late autumn there are frequent but brief dry periods.

6 Grass Steppes of the Southern Hemisphere

Compared with the area of the grass steppes of the Northern Hemisphere, the area occupied by those of the Southern Hemisphere is relatively small. The largest continuous area is the pampa in the eastern Argentinian province of Buenos Aires and parts of the neighboring provinces. The tussock grassland area in Otago on the South Island of New Zealand is quite small.

The pampa lies between 32° and 38° S, extends over about 0.5 million km², and borders directly on the Atlantic coast. It is thus situated in a warm temperate region and corresponds to the southernmost parts of the prairies of Oklahoma and Texas. Rainfall reaches 1,000 mm in the northwest of the pampa and diminishes to 500 mm in the southwest, at its dry limit. Although these values appear high at first sight it has to be remembered that the temperatures, and thus the potential evaporation, are also very high (Buenos Aires mean annual temp. = 16.1° C). Despite this, the climate of the pampa has always been considered to be humid. The question has continually arisen as to why the pampa is bare of trees. The simplest explanation, as in all cases where no other explanation is available, is that the vegetation is of anthropogenic origin, having arisen from an earlier forest vegetation as a result of fires set by man.

This assessment of the climate, has, however, proved incorrect. Even in the wettest parts of the extremely flat pampa, many shallow lakes with no outflow (locally known as lagoons) are present, besides innumerable small pans which, although they contain water in spring, are dried out in summer. The water in the lagoons contains soda and is strongly alkaline. The soils surrounding the pans are alkaline (solo-netz) and support the grass typical of saline soils *(Distichlis)*. All of this points to a semiarid climate such as has already been encountered in the forest-steppe. Measurements of potential evaporation (tank A) showed that in the coastal regions of the La Plata, evaporation and rainfall are equal, but in the pampa evaporation exceeds the rainfall. In the more humid parts of the pampa the negative water balance amounts to 100 mm, and in its arid parts reaches 700 mm. In January and February rainfall is at a minimum and the potential evaporation is particularly high because daytime radiation is very intense and only in the evenings and at night do severe thunderstorms occur. Although well-supplied with water in the spring, the vegetation is severely scorched by January.

On well-drained ground in a forest-steppe climate a woodland vegetation is to be expected and, in fact, in the vicinity of the coast small groves of *Celtis spinosa* (Tala) do occur on slight elevations with a porous limestone or sandy soil. On poorly drained ground there is a grassy vegetation. Almost nothing remains, however, of the original vegetation of the pampa: on the grazed areas European gras-ses have been introduced. They are softer than the pampa grasses and are preferred by the European breeds of cattle. Judging from small remnants occurring on ungrazed patches, it can be concluded that in the humid northeastern portion of the pampa, *Stipa neesiana—Bothriochloa laguroides—steppe* composed of about 23 graminids and 46 herbs originally prevailed. The soil profile beneath such a pampa has a thick humus horizon (1.5 m) and is reminiscent of the thick Chernozems or the prairie soils. There are signs, however, of alternate high and low water content, and the soils are a transitional form leading on to the subtropical grassland soils of southern Brazil. There is no indication of forests having existed previously. Where the ground-water table is high there are stands of the dense tussocks of

Paspalum quadrifarium, which, at a very high ground-water table are replaced by Distichlis on alkaline soils (pH 8—9).

The dry southwestern pampa was previously tussock grassland with *Stipa brachychaeta* and *St. trichotoma,* and was almost entirely lacking in herbs. "Tussock" indicates a growth form completely lacking in the Northern Hemisphere although in the Southern Hemisphere with its mild winters it is widespread. It consists of bunch-like

Fig. 67. Southern tussock pampa with *Stipa brachychaeta* (central province of Buenos Aires).

tufts sometimes more than a meter high in which the hard, old, withered leaves are intermingled with the fresh young greeen leaves, thus providing the tussock grassland with its perpetual yellowish color (Fig. 67). These grasses are of little grazing value, and for this reason they are often ploughed in to give the European grasses a better chance to establish themselves.

Toward the west where the rainfall has decreased to 500 mm annually, falling mainly in the summer, and where the loess soil is replaced by light sandy soil, the pampa is replaced by xerophytic *Prosopis caldenia* woodland. As the rainfall decreases even further, the woodlands are succeeded by *Prosopis* savanna, which strongly recalls the *Acacia* savanna of Southwest Africa (Fig. 68). At the same time large stretches of saline soil are found, bearing a halophytic vegetation. At a rainfall of less than 200 mm annually there is a *Larrea* semidesert on stony ground, with many broom-like bushes belonging to various families (Caesalpinaceae, Scrophulariaceae, Capparidaceae, Compositae). With such a small transpiring surface, and

Fig. 68. Tree savanna with *Prosopis caldenia* and a grass cover of *Stipa tenuissima* and *S. gynerioides* between Sta Rosa and Victoria (Argentine).

by drastically cutting down its transpiration during the six months of drought, the semidesert vegetation is able to survive with the meager amount of soil water at its disposal. This amounts to 50–80 mm annually on flat ground, 25–55 mm on sloping ground, and more than 140 mm in small valleys into which water drains.

The *Larrea* semidesert extends along the eastern foot of the Andes to the northern part of Patagonia, where, south of 40° S, strong west winds blow continuously across the Andes, which at this point are rather low (pass altitude 1,000 m). The wind is, however, of a foehn character, descending and dry. Whereas the eastern margins of the mountains have a rainfall of 4,000 mm annually and support *Nothofagus* forests, these are succeeded to the east by dry *Austrocedrus* forests and, following on these, by a bushland of beautiful red-blossomed Proteaceae *Embotrium coccineum*. The woody plants then disappear and the Patagonian steppe commences. Only 100 km from the Andes the annual rainfall is 300 mm and diminishes even further to 160 mm. Apart from the true steppeland on the westernmost margins of Patagonia, where low tussock grasses are predominant *(Stipa* and *Festuca)*, it is more correct to speak of Patagonian semidesert,

Fig. 69. Patagonian semi-desert with cushions of *Chuquiraga aurea* near Manuel Choique (province of Rio Negro).

characterized by xerophytic cushion plants belonging to completely different families (Compositae, Umbelliferae, Verbenaceae, Rubiaceae, etc.) (Fig. 69). The ground is in many places 60–70 percent bare. The cushion-like form appears to be an adaptation to the constant strong wind (mean velocity 4–5 m/sec); within the cushions a propitious microclimate can be achieved, protected from the effects of the wind.

The Patagonian tussock grassland has much in common with that of Otago on the South Island of New Zealand, situated in the lee of the New Zealand Alps with a rainfall of 300 mm: Both lie to the south of a latitude of 40° S. Low tussock grasses predominate *(Festu-*

ca nova-zelandiae), but at an altitude of 750—2,000 m, where snow remains for 2–3 months of the year, they are replaced by taller tussock grasses, 1.5–2 m high *(Chionochloa = Danthonia)*. Fire and grazing are responsible for the fact that tussock grassland has spread widely in places at the expense of the original Nothofagus forests. So far, no ecophysiological investigations of these grasslands have been undertaken.

7 Semidesert Transitional Zone

Semidesert is distinguishable from true desert by its diffuse vegetation, although the ground is only covered to about 25 percent. In the true desert the density of the vegetation is still lower, and at the same time a change from a diffuse to a contracted vegetation takes place. The plant cover of the semideserts differs greatly. In the frost-free subtropics and in the tropics it consists mainly of woody plants and succulents, in the temperate zone with cold winters mainly of half-shrubs, especially of the genus *Artemisia*. This holds true for the semideserts of Eurasia as well as North America. The characteristic cushion plants of windy Patagonia have already beeen mentioned.

Saline soils are widespread, as would be expected from the greater aridity of the semidesert. This is particularly marked in eastern Europe, where the broad expanses of the "Sivash" to the north of the Crimea dry out in summer and are covered by a salt crust. The salt dust is blown north by the wind and deposited in the southern chernozem and chestnut soil zones, and causes solonization of the soil. In spring the salt is washed out of the upper soil layers by water from the melting snows and the sodium-humus sol thus formed carries the sesquioxides (Fe_2O_3, Al_2O_3) along with it into the deeper soil layers. Here precipitation occurs, and a compact B horizon of strongly alkaline reaction is formed (Fig. 70). The amount of salt deposited increases steadily toward the south.

Humic material is entirely leached out of the A horizon, and the strongly alkaline B horizon becomes harder and harder and, owing to the alternate swelling in the humid season and shrinkage in summer it assumes a columnar structure. This so-called "columnar solonetz" resembles the podsols in certain respects, although the latter are strongly acid in reaction, their peptization being effected by H ions

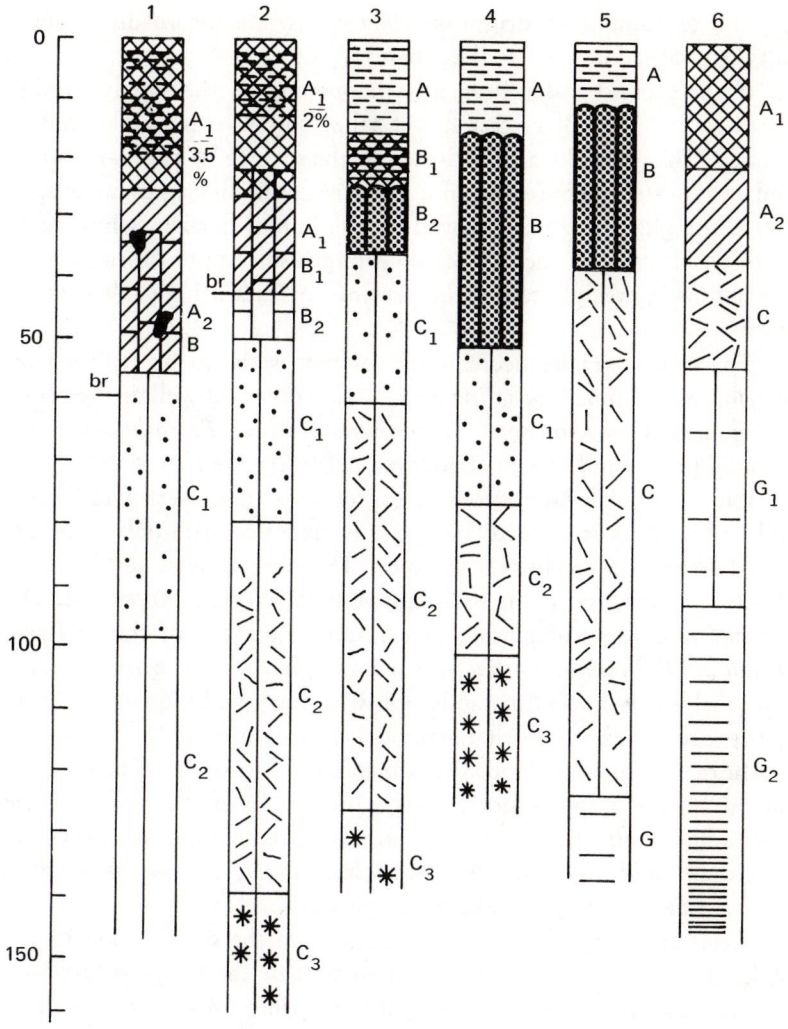

Fig. 70. Soil profiles in eastern Europe, weakly to strongly saline. 1 slightly solonized southern chernozem, slight compaction (A_2B), 2 dark chestnut-soil with B-horizon, 3 light chestnut soil, strongly solonized (A poor in humus and laminated, B columnar and very compact), 4 typical columnar-solonetz soil, 5 solonetz changed by rising ground water, 6 typical solonchak with high ground water and A_1 rich in humus. $CaCO_3$ nodulus C_1, gypsum tubules C_2 in 2—4, and C in 5—6, gypsum druses C_3, gley horizon (ground water) G, G_1 and G_2.

(see p. 195). Beneath the B horizon of the solonetz soils the very slightly soluble $CaCO_3$ precipitates as chalk nodules, followed by gypsum as tubular or druse-like deposits, while the readily soluble salts are washed down into the ground water.

If the ground water rises, as is happening on the slowly sinking north coast of the Black Sea, a wet saline soil known as solonchak is formed. The ground water is drawn to the surface by capillary forces and evaporates. This results in a horizon containing gypsum tubules above the gley horizon, followed by the humus horizon which bears a white salt crust in the dry season. Humus sols are not formed since the humus is precipitated in the presence of such a high salt concentration.

The steppe grasses decrease on solonetz soils, to be replaced by *Artemisia maritima* ssp. *salina* and *A. pauciflora*, as well as species of the genera *Camphorosma, Limonium, Kochia, Petrosimonia*, etc. Ground lichens and species of *Riccia* and *Nostoc* are also found.

On slightly elevated, nonsaline ground a semidesert arid brown soil or burozem, is formed. Its upper horizons contain only 2–3 percent humus and are brown in color. The effervescence level is at a depth of 25 cm, and plant cover amounts to less than 50 percent. The vegetation consists of *Festuca sulcata* and the low half-shrubs *Pyrethrum achilleifolium, Kochia prostata* and *Artemisia maritima* ssp. *incana*, which avoids saline soils. Only solitary individuals of *Stipa* species are seen, but in spring many ephemerals appear.

In the Caspian lowlands the two communities often form a mosaic on the burozem and solonetz soils, resulting from the nature of the microrelief. *Salicornia* and *Halocnemum* predominate on very wet solonchak, and *Suaeda, Obione, Petrosimonia, Limonium caspica, Atriplex verrucifera*, etc. where it is less wet.

After the Caspian Sea had receded from the delta region of the Volga-Ural river system, the southern part of the Caspian lowlands was left covered with alluvial sand upon which *Artemisia maritima* ssp. *incana, Agropyrum cristatum, Festuca sulcata, Koeleria glauca*, etc. then grew. But the vegetation was destroyed by grazing, the sand became mobile, and large bare wandering dunes, or barchanes, were formed. Whenever the sand becomes more stationary, a pioneer vegetation consisting of *Elymus giganteus* and *Agriophyllum arenarium*, a Chenopod, can gain a foothold, followed by species of *Salsola* and *Corispermum*. In the dune valleys *Aristida pennata* and *Artemisia scoparia* among others, make their appearance, and gradually the zonal vegetation is restored.

Sand dunes, particularly those devoid of vegetation, store water. Ground water is always present beneath the dunes, and this leads to the formation of small freshwater ponds in the dune valleys, around which *Elaeagnus angustifolia,* willows, and poplar grow. Attempts to get willow *(Salix acuminata)* and poplar to grow on the sandy areas were initially successful. The plants developed well for the first four years at the expense of the water stored in the ground, but when this was exhausted they died.

In Kazakhstan there are large areas of semidesert between the southern Siberian steppe in the north and the desert in the south. It is comparable to the sagebrush zone of North America, with *Artemisia tridentata* (see p. 128).

8 Central Asiatic Desert with Cold Winters

This region lies to the north of the limit of date cultivation. In the Russian literature a distinction is made between the Middle Asiatic and the Central Asiatic desert (Fig. 71). The former com-

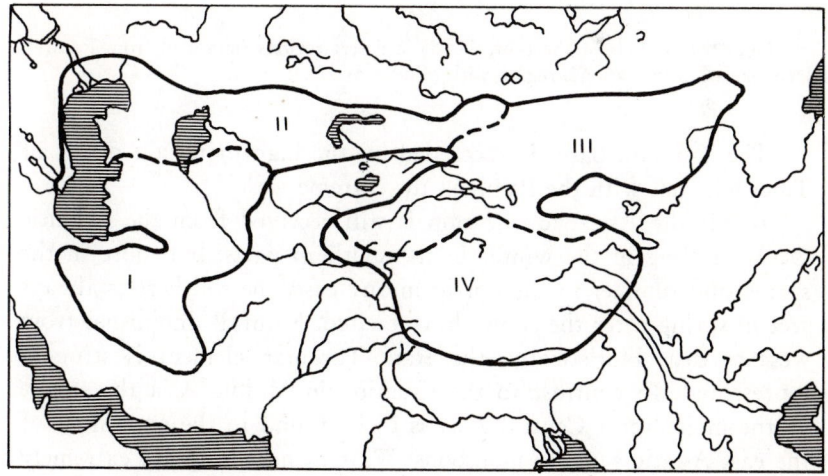

Fig. 71. Asiatic deserts of the temperate climatic zone (from Petrov). Middle-Asiatic deserts: I Irano-Turanian (in parts almost subtropical) and II Kazakhstan-Dzungarian. Central-Asiatic deserts: III in strict sense (hot summer) and IV Tibetan cold high-mountainous desert.

prises the Irano-Turanian desert region occupying the southern portion of the Aralo-Caspian desert and the southern part of Kazakhstan including Dsungaria. The Central Asiatic desert comprises part of Dsungaria, the Gobi desert, the western part of Ordos on the great bend of the Hwang-Ho, Ala-Schan, Bei-Schan, the Tarim basin (Kaschgaria) together with the Takla-Makan desert and the more elevated Tsaidam basin (Fig. 72).

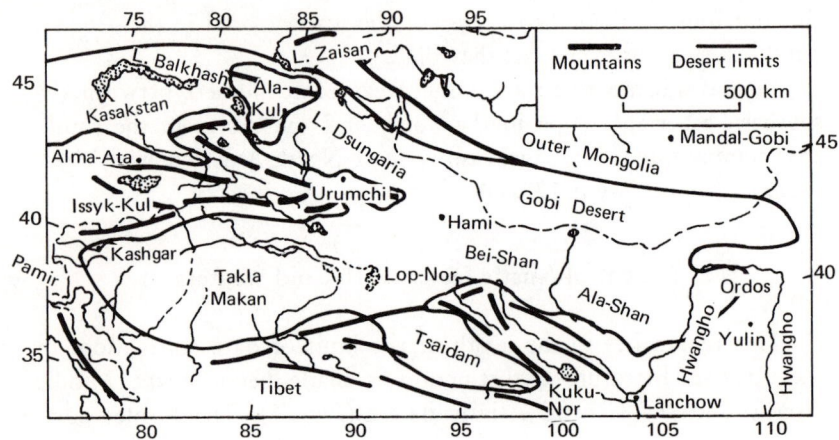

Fig. 72. Divisions of the central-Asiatic desert regions (modified from Petrov). Dzungaria is a transitional region with Middle Asia.

The Tsaidam basin is succeeded by the high mountain desert of Tibet together with the Pamir in the extreme west.

In Middle Asia cyclonic rain is still received from the Atlantic Ocean, falling in the winter in the southern parts, but more in the spring and summer in the north; in any case, the soil here is always wet in spring after the snows have melted. Rainfall diminishes from west to east. Floristically, the Irano-Turanian element is strongly represented. In contrast to the situation in Middle Asia the source of the moisture in Central Asia is to be found in the extensions of the east Asiatic summer monsoons. Winter and spring are extremely dry, and the aberrant rainfall distribution accounts for the predominance of east Chinese elements in the flora (see Fig. 37, Denkoi).

In these regions the most detailed ecological investigations have been carried out on the vegetation of the Middle Asiatic desert in the Aralo-Caspian lowlands (formerly Turkestan). The rainfall in the entire region amounts to less than 250 mm. Since the winters are cold, evaporation at this time of year is very small. This is the reason why the annual evaporation at the Bay of Bogaz is only 1,100 mm. The various types of vegetation are determined by the soils:

1. *Ephemeral desert.* This is found on loess-like, salt-free soils that are very wet in spring but dry from May onward. During the brief period of vegetation, lasting from the beginning of March until May, annual species and geophytes develop, the most common of

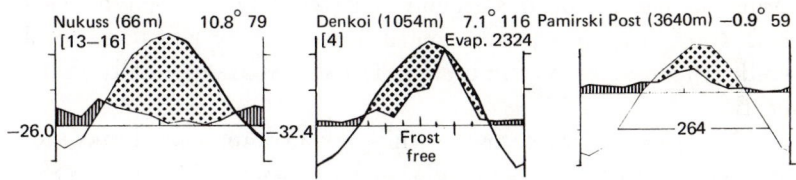

Fig. 73. Climatic diagrams of Nukuss in Middle Asia with winter rain, Denkoi in Central Asia with summer rain and Pamirski Post in the cold desert.

which are *Carex hostii (C. stenophylla)* and *Poa bulbosa.* Here and there the 2-m high *Ferula foetida* is encountered. The 40–50 annual species manage to produce ripe seeds within 30–45 days. In years with a good rainfall the desert presents the appearance of a meadow, producing a dry mass of 0.5 to 2.5 t/ha. It provides grazing for 3 months but is completely lifeless for the rest of the year. It is therefore known as the "starvation steppe".

2. *Gypsum desert.* This is stony desert (Hamada) on the high plateaus of table mountains (mesas). The soil contains up to 50 percent gypsum which has the property of storing water. The situation is similar to that in the Sahara. Therophytes develop in the spring, but otherwise it is the gypsum plants that provide a ground cover (0.1 percent), except in the erosion gulleys where they are more abundant. There are also a few halophytes.

3. *Halophyte desert.* This type of desert is found more extensively, on soils with ground water close to the surface in hollows (Shory), or around salt lakes. Most of the plants are hygrohalophytes *(Salicornia, Halocnemum, Haloxylon, Seidlitzia,* etc.).

4. *Takyrs.* These are bare, clayey, flat expanses which are flooded in spring by surface water running down from the mountains, but which soon dry out again. The shallow pools left behind warm up rapidly and harbor 92 Cyanophyta, 38 Chlorophyta and other algae, producing in all 0.5 t/ha dry matter with an N-content of 4.5% (N fixed from the air by Cyanophyta). Lichens *(Diploschistis,* etc.) colonize slightly higher areas, but flowering plants are rare.

4. *Sandy deserts.* These are particularly widespread: Kara-Kum (black sand) between the Caspian and the Amu-Darya River, Kysyl-Kum (red sand) between the Amu-Darya and Syr-Darya Rivers. The sandy soil favors the growth of a denser vegetation (p. 101). In spring 50 percent of the ground is covered by therophytes (with much *Carex physodes),* to be joined in summer by herbs and shrubs which remain active until autumn. The shrubs are especially characteristic.

Haloxylon persicum, or the white saxaul (pronounced saks-saul), achieves a height of 3 to 4 m and a stem diameter of 35 cm. *H. aphyllum* (black/saxaul) on moist, slightly saline soil is even larger. Besides these, there are the broom shrubs *Calligonum* (Polygonaceae), represented by 30 species, the Leguminosae *Eremosparton, Ammodendron* and *Astragalus* (20 species) and three species of *Salsola.* These plants are either leafless or lose their leaves when water is scarce, and they grow mostly on the ridges of the dunes where the scanty rainwater is stored. The relatively abundant vegetation (an average of 400 kg/ha dry wt) serves as food for the fauna. In the springtime tortoises are a common sight although they spend the remaining 10 months of the year asleep in the sand. In places, the vegetation is ruined by the innumerable rodents. In the past, herds of gazelle, Saiga antelope, wild horses, and donkeys numbering up to a thousand animals were to be seen, but now they have been replaced by flocks of Karakul sheep which have destroyed the vegetation around the water holes to such an extent that here, too, large areas of wandering dunes have been formed.

The largest river of Middle Asia, the Amu Darya (Oxus), is fed by water from the glaciers and snows of the high mountains and is noted for its summer floods. It is accompanied by sedimentation areas with sparse vegetation (Kaire), followed by a zone of tall grass *(Saccharum spontaneum, Erianthus, Typha),* and tall perennial herbs *(Glycyrrhiza)* (Djangil), and flood-plain forests (Tugai). The latter grow to a height of 13–15 m, have a wood reserve of 90–200 m³/ha and consist of *Populus diversifolia, P. pruinosa, Salix, Elaeagnus,* and the lianas *Clematis orientalis, Cynanchum sibiricum,* and *Asparagus persicus.* If the soil becomes saline then species of *Tamarix* and *Halimodendron* (Leguminoseae) can be found.

Numerous ecophysiological investigations have been carried out at the biological station of Repetek (Kysyl Kum), in the Soviet Union. It has been shown, as was to be expected, that plants growing in the sandy desert transpire intensively until the autumn, when a rising cell-sap concentration indicates the onset of a scarcity of water. This explains the relatively large production of organic mass, for which values are also available. Detailed investigations of the composition of the ash from halophytes revealed that there are marked differences between the chloride, sulphate, and alkali halophytes, and the salt-excreting species.

9 Central Asiatic Deserts

As already mentioned, the last traces of the Chinese monsoons are still noticeable in this region, which explains why the rains fall in summer and that the rainfall diminishes from east (Ordos 250 mm) to west (Lop-Nor depression 11 mm). The winter and spring are dry and the spring ephemerals, so typical of Middle Asia, are entirely absent in Central Asia. The flora is poor, the shrubby psammophytes *(Caragana, Hedysarum, Artemisia,* etc.) predominating among the east Chinese-Mongolian elements. *Stipa,* too, is represented by Central Asiatic species. Buckthorn *(Hippophae rhamnoides)* and the tall grass Chii *(Lasiagrostis splendens)* are widespread. Apart from *Populus diversifolia* and *Elaeagnus* the gallery forests contain *Ulmus pumila.* Among the halophytes *Nitraria schoberi* and species of *Zygophyllum, Reaumeria, Kalidium,* and *Lycium* deserve mention.

The character of the deserts is influenced by its geological structure and the nature of the rock:

1. *Ordos.* This is the region lying in the bend of the Hwang-Ho to the north of the Great Wall of China, which runs along the edge of the wandering-dune region. It joins up with the steppe region of the loess plains of the upper Hwang-Ho, nowadays cultivated and dissected by erosion gulleys. The Ordos is a *Stipa* steppe differing

greatly, however, from that of Eastern Europe because of the dry spring. The underlying rock of the true Ordos region is soft sandstone which has given rise to large expanses of sand and dunes with widespread *Artemisia ordosica* semidesert vegetation with *Pycnostelma* (Asclepiadaceae) (cover = 30 to 40 percent). In the central undrained parts there are lakes containing Na_2CO_3 and NaCl.

2. *Ala-Schan.* This is a desert consisting largely of sandy wastes with barchanes. It lies to the west of the Hwang-Ho and stretches as far as the Njan-Schan mountains in the south. To the north it borders the Gobi desert near the Gushun-Nor. Rainfall decreases from 219 mm in the east to 68 mm in the west, while the potential evaporation rises from 2,400 mm to 3,700 mm [1]. The rainfall maximum occurs in August, the mean annual temperature is 8° C, and the minima range from $-20°$ to $-30°$. Ground water is present in the dune region. The encircling mountains have a higher rainfall. Above the desert and steppe altitudinal belts mesophytic shrubs appear at an altitude of 1,900 to 2,500 m; these include *Lonicera, Rosa, Rhamnus,* and *Potentilla fruticosa.* Above this and up to 3,000 m there is coniferous forest with *Picea asperata, Pinus tabulaeformis,* and *Juniperus rigida,* succeeded by subalpine shrubs and alpine mats.

3. *Bei-Schan.* This is the region west of Ala-Schan and is an ancient elevated block rising from 1,000 m to 2,791 m above sea level. It is bounded on the west by the Lop-Nor depression and Hami. Rainfall amounts to 39–85 mm, potential evaporation to 3,000 mm. The vegetation is low shrub-desert consisting of Central Asiatic species with a few halophytes. *Picea asperata* is found growing on the highest points.

4. *Tarim Basin and Takla-Makan.* The basin is 300 km long and 500 km wide, and is surrounded on three sides by high, snow-covered mountains. It is the most arid part of Central Asia, with hot summers and cold winters (minimum $-27.6°$ C). Despite this, it is well provided with ground water fed by the mountain rivers. On an average, the 200-km-long Tarim river carries 1,200 m³/sec and forms wide flood plains. In its lower reaches the river continually changes its course and its water seeps far into the central sandy desert. Lop-Nor is sometimes a salt lake of 100 km diameter and sometimes completely dried out. The sandy desert Takla-Makan is devoid of vegetation,

---

[1] The high potential evaporation probably represents an oasis effect, but the isolated plants in the deserts are affected by it, too.

but water can readily be obtained by sinking wells in the dune valleys.

5. *Tsaidam*. This is an elevated basin at an altitude of 2,700 to 3,000 m, completely surrounded by much higher mountains from which it receives its water. It is cut off from the Lop-Nor depression by the Altyntag. The mean annual temperature is approximately 0° C, the minimum being below – 30° C. The central part of the basin was a large lake in the Pleistocene Age but is today a barren salt desert. *Artemisia* semidesert is found on the sandy soil at the foot of the mountains.

6. *Gobi* (Mongol-desert). This region is north of the above-mentioned deserts and covers the entire southern part of Outer Mongolia. It is separated from the forests and steppes to the east by the Chingan mountains. In the west it touches on Dsungaria which, thanks to rain originating in Atlantic cyclones, is Middle Asiatic in character. The Gobi is gradually replaced to the north by the Mongolian *Stipa*-steppe with *Aneurolepidium (Agropyron)*, and *Artemisia* species. Saline and gypsum soils are common in the desert. The central areas are devoid of vegetation and covered by a stony pavement, and even elsewhere the plant cover is sparse, with a dry-mass production of scarcely 100–200 kg/ha, as compared with 400–500 kg/ha in the northern steppe areas. On low, brackish ground *Nitraria sibirica*, *Lasiagrotis, Kalidium*, etc. are found, and on areas covered by drift sand the saxaul *Haloxylon ammodendron* is to be seen. Nowhere in the entire western Gobi does ground water come to the surface, and there are no oases. The Mongolian Altai mountain range extends into the Gobi from the northwest, continuing as the Gobi Altai. In the latter, only a steppe altitudinal belt is reached, whereas in the former a coniferous belt is present on northern slopes, albeit completely Siberian in character with *Larix*.

10 Cold, High Mountainous Desert of Asia

Lying between the high mountain barrier of the Himalayas to the south and the Kwen-Lun and Altyntag to the north is Tibet, the largest highland mass of the world, with an average height of 4,200 to 4,800 m above sea level. It extends 2,000 km from east to west and is 1,200 km wide from north to south and consists of debris-filled basins, encircled in turn by mountain ranges which are 1,000 m higher still. Water from melting snow forms swampy, frost-debris areas with the Cyperaceous *Kobresia tibetica,* and there are occasional "salt lakes" and even sand dunes.

The monsoon still exerts an influence in the southern and eastern parts, and in the deeply incised valleys forming the upper reaches of the large southern and eastern Asiatic river systems, southeast Chinese and Himalayan forest elements appear.

The larger western and central area, the Changtan desert, is characterized by the most extreme type of climate. The annual mean temperature is $-5°$ C, and only July has a positive mean of $+8°$ C. Daily temperature variations of as much as $37°$ C can òccur, but the rainfall seldom exceeds 100 mm. The flora, which is poor, is very young, having developed after the Ice Age. It consists of Central Asiatic elements *(Eurotia, Kochia, Reaumuria, Rheum, Ephedra, Tanacetum, Myricaria,* etc.). At the western end of the highland plateau, from which the high mountain ranges originate, is the Pamir. At the Pamir Biological Station, 3,864 m above sea level, many Russian scientists have carried out ecophysiological investigations. The mean annual rainfall at this point, most of it falling between May and August, totals 66 mm. The air is dry and solar radiation totals 90 percent of the solar constant so that the ground warms up to $52°$ C in summer, although there are only 10 to 30 nights in the entire year when there is no frost (see Fig. 73, Pamirski Post).

Dwarf shrubs, 10 to 15 cm high, grow on the desert-like habitats: *Eurotia ceratoides, Artemisia skorniakovii,* or *Tanacetum pamiricum.* Along the streams in the valleys, however, there are alpine meadows.

Growth in the dry habitats is extremely slow. *Eurotia* flowers only after 25 years but lives for 100 to 300 years. The root systems are strongly developed, and their mass is 10 to 12 times that of the shoot system. Most of the roots are found in the uppermost 40 cm of the soil, that is to say, in the layers which warm up to more than $10°$ C in summer. Laterally, the roots extend more than 2 m. The reserves of water in the upper meter of the skeletal soil amount to 26 mm at the most and 5 mm at the least. This is very little, but it nevertheless suffices for the scanty vegetation even if the transpiration rate is quite high. Photosynthesis is intense only during the morning hours, and the daily production is given as 25 mg per dm^2 leaf area. Respiratory losses are slowed by the low night temperatures.

It is impossible here to go into the very complicated sequence of the altitudinal belts in the Himalayan mountains, arising as they do in the tropical Indian region and descending on the northern side to Tibet and thus involving enormous climatic differences.

VIII The Boreal Coniferous Forest Zone

1 Boreo-Nemoral Transitional Zone

Unlike the zones discussed so far, the boreal coniferous forest zone of the Northern Hemisphere, with its cold temperate climate, encircles the entire globe through northern Eurasia and North America. To the south, where an oceanic climate prevails, the boreal zone borders on the nemoral deciduous forest zone, but in the regions with a continental climate it adjoins arid steppes or semidesert. There is no sharp boundary between deciduous forest zone and coniferous forest zone; instead, a transitional boreo-nemoral zone is intercalated between the two (Figs. 74, 75). This consists of either mixed stands of a few coniferous species (mainly pine) and a few deciduous species, or of a macromosaic-like arrangement with pure deciduous forest on favorable habitats with good soil, and pure coniferous forest on less favorable habitats with poor soils. In eastern North America different species of *Pinus* represent the conifers in the mixed stands, mainly *Pinus strobus* in the neighborhood of the Great Lakes, although *Tsuga canadensis* is also found, and *Juniperus virginiana* in the Southeast. Pine trees are often the pioneer woody species following forest fires or an abandoned arable land. Since the pines grow more rapidly on poor soils than do deciduous species they constitute the upper tree stratum, but their regeneration in such mixed stands is problematic if there is a dense deciduous undergrowth. For this reason pine trees are only successful where fire plays a recurrent role. It has been shown that fires caused by lightning are a common occurrence in such forests, particularly on sandy soils where there is a dry litter layer in summer.

In Europe the situation is much simpler. On the poor fluvioglacial sands extending as wide belts in front of the end-moraines in Central and Eastern Europe pure pine forests can be found (Pine-

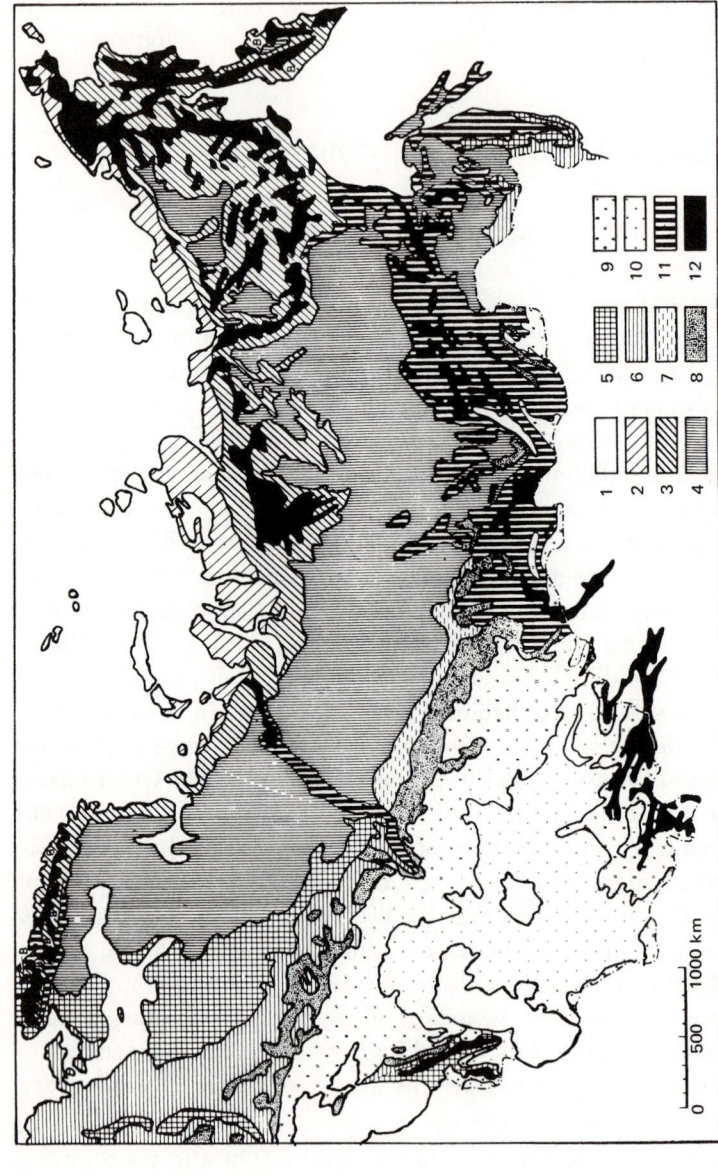

Fig. 74. Vegetational zones of Euro-Siberia. 1 Arctic desert, 2 tundra, 3 dwarf-shrub- and forest tundra, 4 boreal coniferous forest zone, 5 mixed-forest zone, 6 decidous-forest zone, 7 small-leaved decidous forests, 8 forest-steppe, 9 grass-steppe, 10 semi-desert and deserts, 11 mountainous coniferous forests, 12 alpine zone.

192

tum) in regions which belong, climatically speaking, to the deciduous forest zone. In Eastern Europe they are called Bor. On rather better, loamy-sandy soils there is an additional lower tree stratum of oak and the forest is then called Subor, or Querceto-Pinetum. On loamy soils hornbeam *(Carpinus betulus)* occurs as well, and the forests, now with three strata, are termed Sugrudki (Carpineto-Querceto-Pinetum). Finally, on loess, there are the zonal deciduous forests

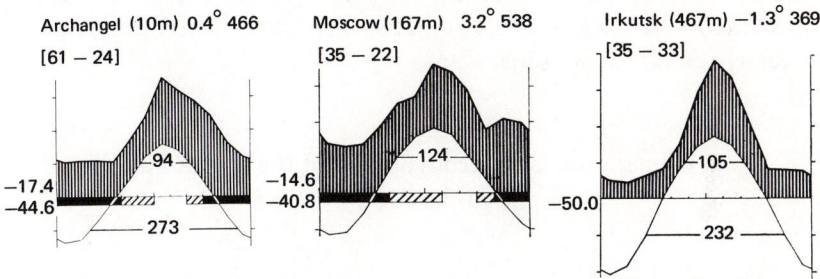

Fig. 75. Climatic diagrams from the boreal zone of northern Europe, the mixed-forest zone and the boreal zone of Siberia.

known as Grud with oak in the upper stratum and hornbeam in the lower stratum (Querceto-Carpinetum). Such forests have been drastically altered by human interference. Forest fires and felling of deciduous trees for fuel have encouraged the growth of pines, whereas the removal of pines for use as valuable building material has resulted in the formation of pure deciduous forests. Yet another disrupting factor is the practice of forest grazing.

In central Europe extensive pine forests have arisen as a result of forestry activities in what were formerly pure deciduous forest regions as, for example, in the Upper Rhine Valley. Further north (in southern Scandinavia and in central Eastern Europe) spruce *(Picea abies)* and oak *(Quercus robur)* are more common. They form a macromosaic but do not mix with each other. Since the better soils which were once the site of oak forests are mostly cultivated nowadays, the proportion of remaining spruce forests has risen. Furthermore, the spruce is encouraged by the forestry industry. In Central Europe the spruce forests at lower altitudes have all been planted by man, and more and more spruce is appearing in the landscape because of its economic value.

The boundary between the boreo-nemoral and the true boreal zone coincides in Europe with the northern distribution limit of the oak. It runs along a latitude of 60° through south Sweden, extends along the southern coast of Finland and thence to the middle Kama River, where the steppe borders on the boreal zone. In the transitional zone between grassy steppe and coniferous forest the aspen, *Populus tremula* in Eurasia and *P. tremuloides* in North America, is very common. In the northern Alps mixed forests of beech *(Fagus sylvatica)* with fir *(Abies alba)* and spruce *(Picea abies)* occur, particularly on calcareous soils.

2 Climate and Coniferous Species of the Boreal Zone
(Fig. 74;4)

The true boreal zone commences at the point where the climate becomes too unfavorable for the broad-leaved deciduous species, that is to say, when the summers become too short and the winters too long. This is recognizable in the climatic diagram as being the point where the duration of the period with a daily average temperature of more than 10° C drops below 120 days and the cold season lasts longer than 6 months (Fig. 75). The northern boundary between the boreal zone and the tundra is where there are only 30 days with a daily mean temperature above 10° C and where the cold season lasts 8 months.

Nevertheless, in view of the large distances over which this zone extends, it would be incorrect to speak of a uniform climate. Rather, a distinction should be made between a cold oceanic climate with a relatively small temperature amplitude and a cold continental climate in which, in extreme cases, a yearly temperature span of 100° C can be registered (from a maximum of + 30° C to a minimum of − 70° C). The floristic composition of the tree stratum in the two types of climate also differs. Like the deciduous forests, the coniferous forests in North America and in eastern Asia contain a large number of different species, whereas those in the Euro-Siberian region contain very few. Many species of the genera *Pinus, Picea, Abies,* and *Larix* as well as *Tsuga, Thuja, Chamaecyparis,* and *Juniperus* are found in North America, although the last four mentioned belong in fact to the transitional zone. The specific representatives of these genera on the Pacific Coast differ from those occurring in the east, and only one species, *Picea glauca*, extends from Newfoundland across to the Bering Straits. Apart from the species mentioned, *Picea mariana*, otherwise usually found on poor soils, is found at the timber line toward the Arctic, and *Larix laricina* in the continental regions. *Abies balsamea, Thuja ocidentalis*, and *Pinus banksiana* also occur, the *Pinus banksiana* on sites previously laid bare by fire. The coniferous belt in the mountain regions contains a large variety of species.

In contrast to this, only two species, the spruce *(Picea abies)* and the pine *(Pinus sylvestris)*, are of any importance in the boreal zone of Europe. Only in the eastern regions is the European spruce replaced by the closely related Siberian species *Picea obovata*, while other species are added to the forests *(Abies sibirica, Larix sibirica*, and *Pinus sibirica*, a subspecies of the Alps-Carpathian *Pinus cembra)*. The proportion of spruce gradually decreases until, in the continental parts of eastern Siberia, it is entirely absent. At the same time *Larix sibirica* is replaced by *L. dahurica*. Larch forests alone cover 2.5 million km² in Siberia. In China to Japan and the nearby parts of eastern Asia the number of coniferous species increases greatly.

3 Forest Types of the European Boreal Forest Zone

Typical of this zone is the dark spruce forest known as Taiga and occurring on podsol soils with a raw humus layer, a bleached eluvial horizon and a compact B-horizon. Soils of this kind are formed from every type of parent rock in the humid boreal zone, but the fewer the bases contained in the rock the better developed is the podsol soil. Litter from spruce (Förna) does not decompose readily and lies above the A_0-horizon, which consists of an organic mass of interwoven rhizomes and roots of the dwarf shrubs, as well as the mycelia of fungi. This is the raw humus layer, and it can readily be removed from the underlying A_1-horizon (mineral soil with humus, which accounts for its being termed superimposed humus or dry peat "Auflage Humus" or "Trockentorf"). Humus acids formed in the humus layer are carried down in the rainwater, completely leaching out the bases and sesquioxides (Fe_2O_3, Al_2O_3) and leaving nothing but fine, bleached quartz sand in the A_2-horizon (Bleichhorizont). At the point where the underlying soil is not yet bleached the humus and sesquioxides are precipitated, either because of decreasing acidity or removal of water by the roots of the trees. This is the origin of the B-horizon, which can be either dark brown (humus podsols) or rusty red (iron podsols). Apart from the tree stratum of the spruce forests (Piceetum typicum), a herbaceous stratum and a closed mossy stratum may also be present. Predominating in the herb layer are bilberries *(Vaccinium myrtillus)*, as well as cranberries in the drier forests *(Vaccinium vitis-idaea)* or, very commonly in the southern zone, wood sorrel *(Oxalis acetosella)*. The following species are also very characteristic: club moss *(Lycopodium annotinum)*, *Maianthemum bifolium, Linneae borealis, Listera cordata, Pyrola (Moneses) uniflora*, etc. Wherever the ground-water table is high more raw humus accumulates, peat is formed, and this in turn leads to the formation of raised bogs, where at first *Polytrichum* dominates in the mossy stratum but is later ousted by peat moss *(Sphagnum)*. If running, well-oxygenated ground water is present, the spruce forests are replaced by flood-plain forests.

Apart from the spruce forests, the proportion of pine forests (Pineta) in the boreal zone is always very high, the pine displacing spruce on dry habitats. The herbaceous stratum of these thin forests consists of heather *(Calluna vulgaris)* and cranberries, as well as such other typical species as *Pyrola* spp., *Goodyera repens, Lycopodium complanatum*, etc. Many lichens are found in the mossy stratum *(Cladonia, Cetraria)*. After a forest fire, which may have been caused by lightning, the burned areas are initially colonized by pine, even though the habitat is normally one favoring spruce. On such burned sites masses of *Molinia coerulea, Calamagrostis epigeios*, or *Pteridium aquilinum* spring up in this order according to increasing dryness of the habitat.

Although birch and aspen are the first to grow on burned sites, they are later ousted by pine, beneath which spruce grows more slowly. In north Sweden the birch stage lasts for 150 years and the pine stage for 500 years, but the fact that fire usually recurs before the zonal vegetation has arrived at the spruce stage explains the high proportion of pine to be found. Only in habitats where there is very little danger of fire is the pine entirely absent.

4 Coniferous Forest as a Biogeocenose

Coniferous and deciduous forest differ in that the former is ever-green. The denser the stand, the less sunshine can penetrate to the forest floor. The soil beneath a spruce forest is 2° colder than in the open. The snow covering, too, is thinner in the forest, so that the ground freezes to a greater depth. The frost depth beneath a dense forest stand in which the frost remained until the beginning of August was found to be 85 cm as compared with 50 cm beneath a thinner stand, from which the frost had disappeared by the beginning of June.

The roots of the spruce are very shallow, usually being confined to the upper 20 cm of the soil or even less if the ground-water table is high. For a high productivity spruce forests require a continuous supply of water and a medium-depth ground-water table, whereas pine, which roots more deeply, is not so sensitive to a dry soil. The total water lost annually by a typical spruce forest amounts to 250 mm in the northern Taiga, 350 mm in the central Taiga, and 450 mm in the southern parts. The mean annual production of organic mass is 5.5 t/ha, and the wood production is 3 t/ha (the latter can reach 5 t/ha in the southern Taiga). The largest annual increment is achieved in the north after 60 years, but in the south after only 30–40 years. The phytomass of the tree stratum of pine forests reaches a maximum of 270 t/ha and that of the undergrowth in old stands 20 t/ha. The quantity of litter produced by older stands on their way to maturity can exceed 1,000 t/ha. This is not accumulated, however, but is continuously decomposed until a state of equilibrium between additions and losses of litter is reached at a litter mass of 50 t/ha. Organic matter accumulates only in the form of peat. Under such unfavorable conditions the annual increase in dry mass of the tree stratum is often less than that of the other strata. In the herbaceous type of spruce-swamp forest the figure is 850 kg/ha for the tree stratum (total 1,906 kg/ha) and for the tree stratum in pine-raised bog, 104 kg/ha (total 1,780 kg/ha). The LAI is relatively high because of the presence on the trees of the needles for at least two years. In pine forests of the boreo-nemoral zone the LAI is 9–10 and in the spruce forests of the Taiga more than 11.

Conifers invariably possess an ectotrophic mycorrhiza, the fungal hyphae greatly enlarging the range of the root system and rendering the nutrients contained in the raw humus layer more easily available to the trees. Plants in the herbaceous stratum are exposed to severe competition from the tree roots. On shallow granitic soil all of the available water may be used up by the pine trees so that a herbaceous stratum is completely lacking and the ground merely covered with lichens. Even the young pine saplings are unable to mature in the face of such root competition and, in fact, only succeed in growing in places where an old tree has died and competition from its roots is therefore lacking. Where the soil is wetter, the roots of the trees utilize the nitrogen in the soil to such an extent that only dwarf shrubs with extremely low nutrient requirements *(Vaccinium myrtillus)* can grow under the trees. However, if the roots of the trees are severed in order to exclude them from competition, conditions of illumination remaining unchanged, other more demanding species take a hold. Examples are provided by *Oxalis acetosella* or even the nitrophilic raspberry *(Rubus idaeus)*, which is otherwise only found in clearings away from the competition of the tree roots. Thus, it is more often the quantity of nutrients available to the plants than the amount of light which determines the composition of the herbaceous stratum.

5 Coniferous Forest Belt of the Central European Mountains and the Upper Forest Limit

Since it is not possible within the scope of this book to deal with all mountainous regions, we shall confine ourselves to a discussion of the situation in Central Europe.

Indigenous to the upper beech forest belt is the white fir *(Abies alba)*, which is ecologically quite different from the Siberian fir. It is strictly confined to the beech regions and occurs as a European montane element from the Pyrenees to the Carpathians and the Balkan mountains.

Above the beech-fir belt in the northern Alps and the Carpathians is the spruce belt, which is completely missing in the Pyrenees and the Apennines. Spruce is definitely continental in its distribution and is particularly widespread in the central Alps, where there is neither beech nor fir. In this region pine *(Pinus sylvestris)* is also common in the lower belts, whereas in the northern Alps it is only to be found in relict habitats in biotopes avoided by other tree species (dolomitic slopes with shallow soil, dry foehn valleys). Three different series of altitudinal belts can be distinguished in the Alps, depending upon the type of climate: that on the northern margins (helvetian), in the central Alps (pennine), and that on the southern margins (insubrien).

Sequences of the altitudinal belts in the Alps:

Helvetian	Pennine	Insubrien
	alpine	
alpine	*Larix-Pinus cembra* forests	alpine
Picea abies forest	*Picea abies* forests	*Fagus sylvatica* forest
Fagus sylvatica forest	*Pinus sylvestris* forest	*Quercus pubescens* forest
Quercus robur forest	(usually absent)	sclerophyllous forest (traces)
(Central European)	(continental)	(sub-Mediterranean)

In the central Alps the uppermost forest belt consists of European larch and *Pinus cembra,* which is related to the Siberian subspecies. The larch plays the part of the light-demanding pioneer species and is, in time, replaced by the five-needled cembra pine, which can better tolerate shade (p. 201). On paths left by avalanches the larch may continue down to lower altitudes. In the Sudetenland and in Poland a special form of larch is found. Above the forest margin occurs "Krummholz" of the two-needled pine *(Pinus montana),* but this is replaced by the shrub-like alder, *Alnus viridis,* on wet habitats.

The northern margins of the Alps have been studied with regard to the factors responsible for setting the upper forest limit. With increasing altitude the period of vegetation is shortened, the summers become cooler, and the winters are both colder and longer. Although these climatic changes take place gradually, the forest limit in high mountains is, in contrast, very sharply drawn. The powers of growth of the trees seem to decrease quite suddenly, and only a very narrow zone consisting of stunted, low forms provides the transition from forest to treeless alpine belt. The question arises as to whether the short summer or the long winter is responsible for the cessation of tree growth, and it would appear that both factors are important. Given a period of vegetation lasting less than three months, the young needles are incapable of maturing properly, and their cuticle cannot attain the required final thickness. As a result of this, during the long winter, and particularly in the strong sunlight of the spring when the ground is still frozen, large water losses occur, as is indicated by rises in cell-sap concentration of up to 65 atm and more. Damage typical of frost-drought is observable, and the needles drop off. Beneath a covering of snow such events cannot take place, and this explains the ability of the stunted forms to survive to some distance beyond the forest limit. It is apparently the combined effect of the two factors, the shortened period of vegetation and the increased danger of frost drought, that is responsible for the abruptness of the forest limit at a certain altitude.

Pinus montana, which extends beyond the limits of the spruce forests, manages with a shorter period of vegetation. But about 100 m further up the phenomenon repeats itself, the needles are unable to mature, suffer damage from frost drought, and the upper *Pinus montana* limit is just as sharply defined as that of the forest.

The factors responsible for setting the limits of polar forests have not been investigated, but they are probably similar, apart from the fact sunshine plays no part in the damage caused by frost drought. This is probably replaced by the drying effect of the strong, cold winds, as is borne out by the observation that in the sheltered valleys the forest limit pushes further to the north than on the watersheds.

At its highest, the forest limit in the central Alps lies at 2,000–2,150 m and is, as already mentioned, formed not by spruce, but by deciduous larch and evergreen cembra pine which has relatively delicate needles. Continuous measurements of climatic factors and of photosynthesis have been made here throughout the entire year so that the productivity of larch and cembra pine could then be compared. During the cold winter photosynthesis is at a standstill, even in the evergreen cembra, but in spring the evergreen needles rapidly come into a state of activity whereas, at these altitudes, the larch is not green until the middle of June and is beginning to turn yellow by the end of September. Not more than 107 days are at its disposal for production, as compared with the 181 days available to the cembra pine. However, the young larch has 3 to 6 times the mass of needles possessed by the young cembra pine, besides which, despite the brief period of vegetation, it assimilates 47 percent more CO_2 per g of needle. The total production, therefore, of a 4-year-old larch is 4.5 times, and that of a 12-year-old, 8.5 times greater than that of cembra of the same age. Only from the 25th year onward is the quantity of needles produced by the larch the smaller of the two, and it begins to lag behind in its growth, particularly on raw humus soils. In time, the cembra succeeds in establishing itself as a shade-enduring species. The relationship of larch to cembra is similar to that of pine to spruce.

6 Ecophysiological Investigations on Coniferous Trees

The following information concerns the water economy of a spruce forest in Sweden:

A large part of the rainwater is witheld by the crowns of the trees (interception), 50 percent in Sweden but only 30 percent in the thinner east European stands. Moss and litter layers retain a further portion of the water and, finally, only about one-third of the total rainfall reaches the roots. In the summer months this was found to amount to 90 mm, and for the rest of the year 202 mm, making a total of 292 mm, a quantity which is almost completely lost by transpiration in a 40-year-old stand. In wet habitats as much as 378 mm are lost to the atmosphere by transpiration, which means that a part of the water must be drawn from the ground water.

Such active transpiration is paralleled by equally intense photosynthesis. The spruce possesses two kinds of needle, sun needles and shade needles. The situation is reminiscent of that in the beech, but with the difference that the active period for the evergreen spruce begins very early in spring and continues into the autumn, until the onset of occasional frost. The seasons of low nocturnal temperatures and small respiratory losses are particularly favorable for the net gain of dry substance. Nevertheless, after a night of frost, photosynthesis is temporarily inhibited although it is not until the beginning of the cold season proper that the spruce falls into a state of dormancy in which it does not even assimilate on sunny days.

At the same time, respiration sinks to such a low level that it can hardly be measured and accounts for only negligible material losses. The needles lose their fresh green color at this time, and the chloroplasts are difficult to recognize under the microscope.

Following a long period of cold it takes a little time for photo-synthesis to regain its normal level in the spring, since the photosynthetic apparatus must be reactivated. Young cembra saplings in the mountains have been shown to spend the winter beneath the snow with green needles and to recommence CO_2 assimilation immediately in the higher temperatures of the spring.

The transition to winter dormancy is accompanied by a process of "hardening," or, in other words, a great increase in resistance to frost (see p. 159). Whereas in their nonhardened autumn condition spruce needles are killed by frost at a temperature of $-7°$ C, they are capable of enduring a temperature of $-40°$ C in the winter without suffering any damage.

The resistance of the needles to frost can be artificially changed, the hardening process in particular, by the influence of low temperatures in late autumn and spring and the dehardening process by the influence of normal room temperatures, especially in December and late winter. Hardening prevents the occurrence of frost damage in coniferous trees in their natural habitats even at temperatures as low as $-60°$ C, such as occur in Siberia. Thanks to the state of winter dormancy they are also in position to survive the complete darkness of the polar winters. The varying degree of adaptation achieved by the different species is reflected in their distribution. Only a few species can tolerate the extremely continental Siberian winter, the deciduous Siberian larch better than the evergreen species. Variations within a species also occur, depending upon their provenance. Spruce from the Alps behaves differently from members of the same species taken from the northern boreal zone or, again, spruce from the upper tree limit behaves differently from spruce from lower altitudes. The more extreme the conditions, the more pointed do the crowns of the trees become, showing that the growth of the lateral twigs is more strongly inhibited than that of the main shoot. In polar regions the same phenomenon is observable in pine.

Whether or not this shape results from a selection of mutants better able to withstand the weight of snow is unknown, but the same phenomenon has been observed in fir trees at the lower, dry limit in Albania, where snow is not important. The most likely explanation is that whenever the general situation is unfavorable the growth of the lateral twigs is inhibited before the main shoot suffers (the reverse is true if light conditions are poor). On dry slopes in Utah in North America, *Picea, Abies,* and *Pseudotsuga* have pointed crowns whereas they are rounded on the valley floor, where water conditions are better.

7 Mires or Bogs of the Boreal Zone

The boreal zone has a humid climate, the rainfall exceeding the potential evaporation, which means that the water balance is positive. If, in any way, the surplus water is prevented form draining into the rivers, the ground-water table rises and mires are formed. Since the soils of the boreal zone are poor and acid (podsols), the ground water is acid in reaction and has a low mineral content. It is usually brown in color due to humus sols. The situation only differs if the underlying rock is limestone. Large expanses of the boreal zone in Euro-Siberia and in North America are very flat so the ground-water table is high. As long as it remains more than 50 cm below the ground for the larger part of the year, tree growth is possible, otherwise this is inhibited and the forests are replaced by mires. Extensive areas of the boreal zone are covered by communities on peaty soils and not by the true zonal vegetation, which is coniferous forest. In large areas of Finland more than 40 percent of the total land is covered by mires, in places even 60 percent. The same holds true for the boreal zone of eastern Europe and especially western Siberia which, except in the vicinity of the rivers, is entirely covered by swamps and mires. In Kamchatka, Alaska, and Labrador, as well as the regions to the south of Hudson Bay, the situation is in places similar. For this reason the mires have to be dealt with after the coniferous forests. The dividing line between the two is often difficult to establish. In the spruce forests already mentioned, with *Polytrichum* and *Sphagnum,* peat formation is well developed.

In the geological meaning of the word, a mire, "Moor" in German, must have a peat layer at least 20—30 cm deep. If the peat layer is thinner or if its content of combustible material falls below 15 – 30 percent then, in German, the term "Anmoor" is applied. In the ecological sense mires are plant communities which are dependent upon a high ground-water table but independent of the thickness of the peat layer upon which they grow.

On account of the poor aeration of the soil the roots of the plants remain near the surface so that only the nature of the uppermost peat layers is of interest to them.

Three types of mires can be distinguished according to the origin of their soil water:

1. *Topogenous mires*. These are associated with a very high ground-water table and for this reason occupy the lowest portions of the relief or occur wherever spring water is available. Many widely differing types of *fen* belong to this group.

2. *Ombrogenous mires* or *raised bogs*. These are higher than their surroundings and are exclusively watered by the rainwater falling onto them.

3. *Soligenous mires*. These are also watered by rain but also receive water draining in from surrounding slopes, because they are not higher than the surrounding country.

The ground water of the topogenous mires or fens may contain mineral substances and be rich in nutrients, if so the mires are termed eutrophic or minerotrophic. Rainwater, on the other hand, is very pure and poor in nutrient substances so that the ombrogenous mires are said to be oligotrophic or ombrotrophic. The run-in water received by the soligenous mires, unless it comes from melting snow, contains rather more nutrients and such mires are usually termed minerotrophic.

Since the ground water of the boreal zone is poor in mineral salts, it is not easy to distinguish between fens and bogs and it is more usual to speak of mesotrophic transitional bogs. If the water contains less than 1 mg Ca per liter the less exacting species typical of the oligotrophic mires are to be found.

Eutrophic mires or fens in which *Carex* spp. play a leading role occur in the temperate zone independently of climate, if the ground water contains calcium but is not saline. This is just as much an azonal vegetation as the swamp plant and aquatic plant communities.

We are concerned here with the oligotrophic mires, which are only to be found in cool to cold climates and of which several distinct types can be recognized, according to their structure and topography (Fig. 76).

1. *Blanket bogs*. These have already been encountered in the extreme oceanic climate of the Atlantic heath region of the British Isles

and along the entire west coast of Scandinavia (p. 150). They cover the entire terrain.

2. *Raised bogs.* These are typical of the rather less oceanic northwest corner of Central Europe with its heath regions, of the entire boreo-nemoral zone and of the southern part of the boreal zone. In their typical form they are devoid of trees, but as the climate becomes drier and more continental, pine grows out into the bogs to form the so-called forest-raised bogs. The entire southern margin of the boreal zone consists of such bog forests (Fig. 76).

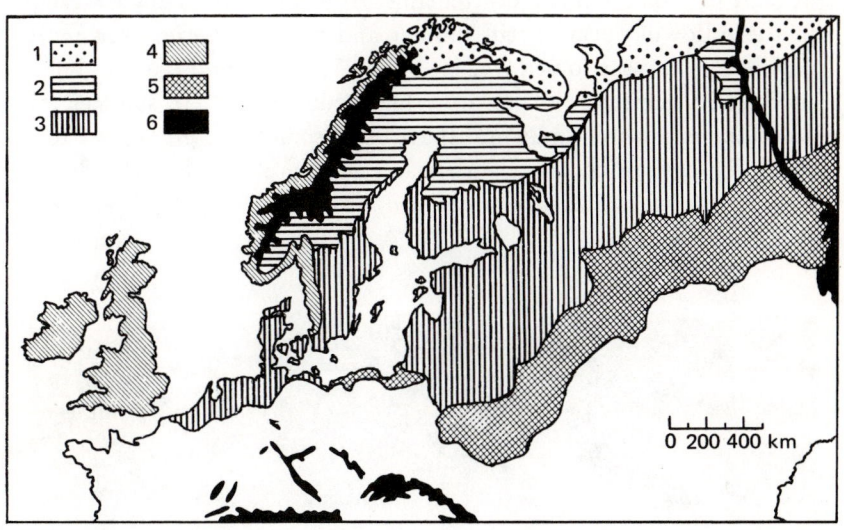

Fig. 76. Distribution of the various types of mires (bogs) in N. Europe (modified from Katz and Eurola). 1. Paalsa bog, 2. Aaapa mires, 3. typical raised bog, 4. "blanket" bog, 5. forest-raised bog, 6. mountain bogs. The white patches in the southern regions have mainly topogenous fens.

3. *The Aapa mires or string bogs.* These are found north of the raised-bog zone, but above all in Fenno-Scandia. They are gently sloping soligenous mires with slightly raised ombrotrophic ridges running at right angles to the slope. In between the ridges are elongated depressions filled with minerotrophic water (Finnish "Rimpis," Swedish "Flarke"). The entire mire presents a descending, terraced aspect reminiscent of terraced rice fields. The ridges are in part the result of the lateral pushing effect of the ice covering the Rimpis in winter.

4. *The Palsa bogs of the peat-hummock tundra.* These begin beyond the boreal zone in the forest tundra, in regions where the annual
mean temperature is below $-1°$ C. Ice in the ground itself is partly
responsible for the formation of the peat hummocks which may be as
much as 20 to 35 m long, 10 to 15 m wide, and 7 m high. On slightly
elevated ground the snow cover is thinner, and the frost can penetrate more rapidly into the peaty soil. Layers of ice are formed which
attract water from the unfrozen surrounding peat. As the ice thickens,
it pushes up the peat, and since not all of the ice melts in summer
at least part of the hummock remains. As a result, the snow covering
in the following year is still thinner and the soil freezes even more
rapidly. From year to year the ice becomes thicker, and the peat hummocks with their core of ice grow higher and higher. In summer the
structure as a whole sinks into the ground somewhat, giving rise to a
ditch-like depression filled with water in which dwarf birch *(Betula
nana)* and cotton grasses *(Eriophorum)* grow (Fig. 77).

Fig. 77. Palsa or peat hummock bog in N. Finland.

The tops of the peat hummocks (Palsen) dry out in summer, crack,
and undergo wind erosion. It can be assumed that the majority of
Palsen are subfossil structures in a state of disintegration.

Whereas the raised bogs of Eurasia and North America are confined to the oceanic regions, the Aapa and Palsa bogs are circum-
polar in distribution.

8 The Ecology of Raised Bogs

Peat mosses *(Sphagnum* spp.) play the largest part in the formation of raised bogs. They contain large numbers of dead cells which

easily fill up with water by capillary action so that the cushion-like plants are of a spongy nature and contain many times their own weight of water. As the plants grow upward, the lower ends die off and are converted into peat. As the cushions grow larger and larger, they merge with one another until finally the entire area presents the watch crystal-like aspect of a typical raised bog. Because Peat mosses, cannot tolerate drying-out, they require uniformly damp, cool summers. Since they colonize only poor, acid soils, the podsols are well-suited to their requirements. For this reason raised bogs usually originate in boreal coniferous forests which have gradually become wetter.

In a large, growing, raised bog it is possible to distinguish a very wet, only slightly convex *high area,* a better drained and relatively steep *marginal slope,* and a surrounding *minerotrophic fen,* termed a "Lagg." The high area is not absolutely flat, but consists of small hummocks, the "Bults," which extend above the mossy areas, and "Schlenken"or small hollows in the mossy carpet that are filled with water and in which such hygrophilic peat plants as *Carex limosa* or *Scheuchzeria* can grow. Several Schlenken may unite to form the bog-pools known as "Blänken" or "Kolke." They are usually 1.5 to 2 m deep and are filled with soft detritus. Surplus water drains off the high areas in small gullys called "Rüllen".

Only a few flowering plants are able to survive in the raised bogs and they have to be species that are undemanding as regards nutrients. Examples are provided by *Eriophorum vaginatum, Trichophorum caespitosum,* besides the dwarf shrubs *Andromeda polifolia, Vaccinium oxycoccus, V. vitis-idaea, V. uliginosum, Calluna vulgaris* and *Empetrum.* In Atlantic regions *Narthecium* is to be found; in the east *Ledum palustre* and *Chamaedaphne calyculata,* and in the north *Rubus chamaemorus, Betula nana,* and *Scheuchzeria palustris.*

The second ecological factor affecting the survival of other plants, apart from the scarcity of nutrient materials is the danger of their being overgrown by the peat mosses. The viable tips of the latter are the substrate upon which the flowering plants have to germinate. Depending upon the amount of water available, the peat mosses grow 3.5 to 10 cm annually, and, if they are not to be smothered by the mosses, the flowering plants are obliged to raise their shoot bases by this amount, either by elongating the rhizome or by putting out adventitious roots. In places where the peat mosses grow more slowly, such as on the relatively dry hummocks or on well-drained slopes, it is easier for other plants to survive, and here dwarf shrubs are usually to be seen. Definite zones can be distinguished on the individual hummocks: *Eriophorum vaginatum* and *Andromeda* at the base, *Vaccinium oxycoccus* further up, and other dwarf shrubs at the top. The top of the hummock is often so dry that *Sphagnum* is replaced by other mosses *(Polytrichum strictum, Entodon schreberi)*, and even by lichens *(Cladonia* spp., *Cetraria).*

Trees confronted with Sphagnum moss are at a great disadvantage since not only is the base of their trunk fixed but their growth is also extremely slow on such a poor substrate. Often not more then the topmost twigs are higher than the hummocks. Bog forest occurs only in places where the growth of the peat mosses is limited by a drier climate. Once a bog is drained peat moss ceases to grow and the area is rapidly converted into heath where dwarf shrubs assume a leading role, soon to be joined by trees such as the birch, pine or spruce. This is the state of the majority of contemporary bogs in Central Europe.

Even though the water is just as poor in nutrients as in the rest of the bog, species typical of minerotrophic soil often grow along the runnels or on the edges of the pools. It appears that running water or water agitated by waves provides the plants with more nutrients than stagnant water, in which a mere diffusion of nutrients takes place. Because of their high water content bog soils warm up very slowly and are therefore cold habitats, which explains the presence of northern Arctic floristic elements including relics of the Glacial period. Furthermore, on a raised bog these relics do not have to compete with more rapidly growing and exacting species.

With the exception of *Drosera* spp., which supplement their nitrogen supplies considerably by digesting the insects caught on their leaves, all of the other plants are xeromorphic, despite there being water to excess at their disposal. This is aseribed to a lack of nitrogen. It has generally been observed that *xeromorphism occurs whenever the growth of the plant is inhibited by a deficiency*, for example by lack of or excess of water (leading to poorly oxygenated soil), by low soil temperatures which hinder the uptake of nitrogen or by a direct nitrogen deficiency. This xeromorphosis is a symptom of deficiency, and it is therefore more accurate to use the term *peinomorphosis* (Greek, peina = hunger).

IX The Arctic Tundra Zone

1 Forest Tundra

Just as the forest-steppe forms a transitional zone between forest and steppe proper, a macromosaic arrangement of forest and tundra provides a transition from the boreal forest zone to the treeless tundra. The first sign of a transition is the occurrence of scattered treeless patches, usually on raised ground, within the forest region. These become more frequent toward the north until only scattered islands of forest remain, finally consisting of low stunted malformed specimens. Whereas this zone of stunted trees is quite narrow in the mountains (p. 199), it may extend for hundreds of kilometers on flat land.

The tree most typical of oceanic regions is the birch, of the extreme continental regions the larch, and elsewhere the spruce. Apart from the factors already mentioned (p. 200), failure to regenerate is held to be one of the factors determining the timber line in polar regions. At their northernmost limits of distribution trees rarely produce seeds capable of germination, and those that they do produce are often eaten by animals or swept by storms far up to the north over the smooth, snowy surfaces to regions where they can no longer develop. The thick blankets of moss and lichens found in forest-tundra provide a very unsuitable substrate for the establishment of tree seedlings.

Man and his accompanying herds of reindeer play a large role in this part of the world, both on account of the damage done by the animals and, even more, as a result of wood utilization, the natural rate of growth of the trees being extremely slow. As a rule, a tree seedling succeeds in establishing itself only if the temperature has been especially favorable for two years in succession. Even so, its further growth is very slow and after 20 to 25 years the tree may scarcely be taller than the plants in the herbaceous layer, the annual increase in height amounting to only 1 to 2 cm. The size of yearly rings of the stem is closely correlated with the July temperatures.

Open areas in the forest-tundra are usually occupied by dwarf-shrub tundra, which also constitutes the southernmost zone of the true tundra (Fig. 74).

2 Climate and Vegetation of the Tundra

The largest tundra region completely devoid of forest is an area of 3 million km² in northern Siberia. At the most there are 188 days in the year with a mean temperature above 0° C and sometimes as few as 55. The low summer temperatures are partially due to the large amount of heat required to melt the snow and thaw out the ground. Winters are rather mild in the oceanic regions but extremely cold in the continental regions (Fig. 78). However, the cold pole still

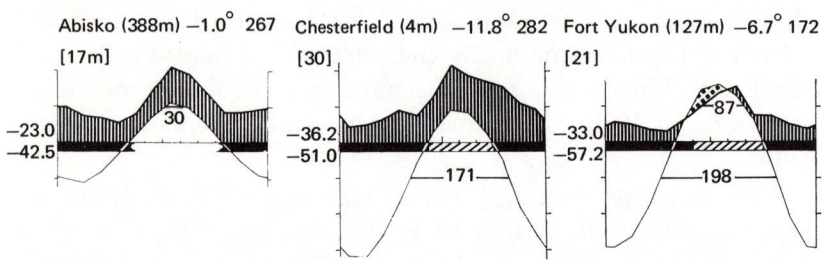

Fig. 78. Climatic diagrams from the forest tundra of Sweden (oceanic), from the tundra of N. Canada and from the extremely continental boreal region of Alaska (see also Fig. 8, Verchojansk).

falls within the forest region, near Verchojansk and Oimekon although the mean annual temperature at this point is − 16.1° C and the permafrost extends far down into the ground. The depth to which the ground freezes in winter has no influence upon the vegetation as plants depend only on the thickness of the upper soil layer thawing out in summer.

In the southern tundra the growing season commences in June and lasts until September. The wind is of great importance here because it causes irregular drifting of snow which, in turn, is responsible for the mosaic-like arrangement of the vegetation. In winter the storms can reach a wind velocity of 15–30 m/sec.

Precipitation is slight, often being less than 200 mm, but since the potential evaporation is also very low the climate is humid. Surplus water is unable to seep into the ground because of the permafrost, and thus extensive swamps are formed. The amount of peat formed, however, is negligible because plant productivity is low. Snowfall amounts to 19–50 cm annually, the raised ground being blown free of snow so that the abrasive action of snow and ice are decisive mechanical factors influencing the vegetation.

The steep, stony slopes that warm up relatively well in summer when the sun is low in the sky, often look like flower gardens. Together with the banks of streams and rivers they constitute the most favorable habitats in this zone. Flat, raised ground with stone nets polygonal soil) and gentle slopes subjected to solifluction (19) are, on the other hand, sparsely colonized. Interminable stretches of land are covered with dwarf birch and dwarf willows, and with *Eriophorum* and *Carex* spp. The drier soils support a pure lichen tundra, whereas mosses predominate on wet ground (no *Sphagnum* spp.). As far as air temperature is concerned, meteorological measurements carried out at a height of 2 m are no indication of the temperature of the low plant cover. By the time the air temperature has risen to 0° C, the ground has usually already thawed out to a depth of 0.5 m and the development of the vegetation is in progress. In fact, the daytime temperature of the plants is often 10° C higher than that of the air, but, in spite of this, the summer is often too short for the seeds to ripen. In an attempt to overcome this problem,half of the plant species in Greenland, for example, produce their flower buds in the preceding season so that no delay is involved in their coming into flower at the beginning of the following warmer season. Buds and leaves usually spend the winter safely beneath a covering of snow, despite temperatures as low as − 30° C, whereas open. flowers would die. Reports that *Cochlearia arctica* survives the winter in flower beneath the snow are based on a misunderstanding.

Of particular interest are the aperiodic species, such as the tiny crucifer *Braya humilis*. Their development is protracted over several years and can be broken off temporarily for the winter at any stage. In this manner they are unaffected by the shortness of the summer and may flower either at the beginning of the vegetational season or at a later date, the buds having been formed as much as two years previously. Dispersal of seeds and fruits in 84 percent of the species is carried out by the wind skating them over the snow and in 10 percent by water; berries are found only in the forest tundra. The size of the seeds reflects the low productivity of the tundra, 75 percent of the species producing seeds weighing less than 1 mg. The majority of plants in the tundra require the influence of the low winter temperatures before they are capable of germination, but in spring they are then in a position to germinate rapidly and have time to accumulate some reserves before the autumn. A variety of grasses, as well as *Polygonum, Stellaria,* and *Cerastium* spp., accounting for 1.5 percent of the total species, are viviparous (pseudoviviparous). Open spaces, such as occur on the lower Lena, can rapidly be colonized because of the extremely prolific seed production. The majority of tundra species are hemicryptophytes or chamaephytes. The brief period of vegetation at low temperatures is unfavorable for annuals or therophytes (in contrast to conditions in the desert), which are therefore represented only by *Koenigia islandica* (Polygonaceae), 3 species of *Gentiana,* by *Montia lamprosperma,* 2 species of *Pedicularis,* and a few others. Most of the plants develop thick roots which serve as storage organs. Individual plants, including herbaceous specimens, may live more than 100 years, the dwarf shrubs reaching an age of 40 to 200 years.

Nitrogen supply presents a problem of great importance in this region. Leguminosae possess root nodules lying immediately beneath the surface, where the soil warms up in summer. If there is no nitrogen in the soil, then only mosses and lichens are to be found: The animal dung is of great value as a source of nitrogen. It has recently been shown that *Dryas drummondii* a pioneer species growing in Alaska, has nodules similar to those of *Alnus,* and that the nitrogen content of the soil rises from 33 kg/ha up to 400 kg/ha during the pioneer stage of its colonization.

In some of the trough valleys in the interior of Peary Land (north Greenland) at a latitude of 80°, the climate is completely different from that of the rest of the Arctic. As a result of the descending winds blowing from the interior in summer, there is no rain and desert-like conditions prevail, with salt efflorescence and an alkaline soil supporting a few halophytes. Apart from these plants, vegetation is not completely lacking because the snow drifting down from the mountains in winter provides a source of water in spring when it melts. Because the soil thaws out to a depth of about 1 m, the water can seep down and is then available for plants such as *Braya purpurescens,* which can develop tap roots over 1 m long. 59 days of the year are frost-free, and the mean July temperature is 6° C.

3 Ecophysiological Investigations

Arctic plants have a well-balanced water budget, and their cell-sap concentration lies between 7 and 20 atm. The fact that some of them exhibit xeromorphic features can be attributed to an inherited peinomorphosis caused by nitrogen deficiency, similar to that seen in plants of raised bogs. A low soil temperature renders the uptake of nitrogen more difficult. Photosynthesis and the resulting production of organic matter are of vital importance. The maximum CO_2 assimilation does not exceed 12 mg/dm²/hr, and on cloudy days the CO_2 uptake curve temporarily sinks below zero. But since photosynthesis can usually be carried on around the clock, with a minimum at midnight when illumination is poor, the uptake of CO_2 on a summer's day amounts to 100 mg/dm²/hr, which corresponds to 60 mg starch.

Such quantities suffice for the accumulation of adequate reserves in the summer. The primary production of the plant cover in one year in the sub-Arctic regions of Swedish Lapland near Abisko (growth season of 111 days) amounts to 2,500 kg/ha, in Alaska

Total Aboveground Phytomass and Primary Production in t/ha

Vegetational Zone (Climatope)	Aboveground Phytomass	Primary Production
Northern tundra (cushion plants)	0.6	0.4
Central tundra (rich in herbs)	1.2	0.7
Southern tundra (dwarf shrubs)	3.2	1.2
Forest-tundra with birch and spruce	7.3	1.4
Northern spruce forest with mosses and lichens	90.0	1.5
Central spruce forest with bilberries	130.0	3.0
Southern spruce forest with wood sorrel	220.0	5.0
Deciduous forest (oak)	260.0	5.6
Meadow-steppe	1.5	1.5
Feather-grass steppe (abundant herbs)	1.2	1.2
Feather-grass steppe (few herbs)	1.0	1.0
Sagebrush steppe	0.5	0.5
Sagebrush semidesert	0.7	0.4

(growth season 70 days), 830 kg/ha, and in the extreme Arctic (60 days), only 30 kg/ha. In an Arctic willow scrub in Greenland the phytomass may reaches 5.5 t/ha.

A few comparative figures for the different vegetational zones of eastern Europe are given in the preceding table by E. M. Lavrenko, W. H. Andrejev and W. L. Leontjev.

4 Animal Life and Plant Cover of the Arctic

The extensive tundra of Siberia provides us with one of the few remaining opportunities of encountering a fauna in its original state and of studying its influence upon the vegetation. The majority of larger vertebrates leave the tundra in winter, the birds migrating to the south, leaving only the lemmings and ground squirrels behind. The polar fox and the snowy owl move down from the northernmost regions where prey has become scarce.

Lemmings neither hibernate nor do they lay up stores for the winter. Rather they remain active beneath the hard covering of snow and exist mainly upon the young buds of the Cyperaceae. Although itself weighing only 50 g, a lemming requires 40 – 50 kg of fresh plant material annually. As a rule it colonizes well-drained southern slopes and builds a nest of Cyperaceae shoots in the vicinity of its feeding area, which, for one family, usually covers 100–200 m². An entire colony occupies about 1–1.5 ha and destroys 90 to 94 percent of the entire vegetation. In such places *Eriophorum angustifolium* does not succeed in flowering. The number of lemmings reaches a maximum on an average every three years. The dried-out vegetation is not eaten but forms a kind of hay in spring (1–1.2 t/ha), which is then washed together and forms peaty hummocks. When they abandon their winter quarters, the lemmings proceed to build on higher ground, throwing up as much as 250 kg of earth per hectare.

On disrupted habitats such as these occurs a characteristic plant community which is the beginning of a secondary succession. The same situation is produced by the burrowing of ground squirrels, and in this manner the plant cover is held in a continuously dynamic state. Flocks of water birds, mainly geese, arriving in spring, destroy 50 to 80 percent of the plant cover by pecking off the young shoots of *Oxytropis* and pulling up the starchy rhizomes of *Eriophorum*.

Either solifluction sets in on the bare ground, or a thick covering of moss develops.

The nesting and flocking places of these birds are well manured, and nitrophilous species such as *Rhodiola, Stellaria, Polemonium, Myosotis, Draba, Papaver,* etc. — grow there.

Although the reindeer remain in the tundra in winter only if large snow-free areas are available for grazing, they, too, must be considered as belonging to these regions. In summer the animals graze singly and have little effect on the vegetation, but in the autumn, when they gather together in large herds, their trampling is very obvious. Lichens and dwarf shrubs are destroyed, so that grassy communities consisting of *Deschampsia* and *Poa* gain ground. The number of wild reindeer is decreasing nowadays in favor of the domesticated reindeer. Animals of prey, such as the polar fox, have only a small direct effect upon the plant cover.

5 The Antarctic and Subantarctic Islands

Only two flowering plants have been found on the edges of the ice-covered Antarctic continent: *Colobanthus crassifolius* (Caryophyllaceae) and a grass, *Deschampsia antarctica.* In recent times *Poa pratensis* has been imported, but otherwise only mosses, terrestrial algae, and lichens are to be found. They are confined to places on the coast that are sometimes free of snow, to steep cliffs, and to talus, but their phytomass is negligible.

The sea surrounding the Antarctic, with its continous westerly storms, is scattered with many tiny islands, most of them south of the 50th parallel. They are bare of trees since the summers are cool, although the winters are not cold. Drizzling rain and fog occur all the year round. Sometimes they are referred to as wind deserts, since only in sheltered places is there a somewhat more luxuriant vegetation.

The most common plant on the Kerguelen Islands is the dense, cushion-shaped *Azorella selago* (Umbelliferae). In earlier times the seafarers were aware of the antiscorbutic action of the large-leaved Kerguelan cabbage *Pringlea antiscorbutica* and used it as a fresh vegetable. *Acaena* species are common on all of the islands and, besides many mosses, ferns, and lichens, tussock grassland *(Festuca-*

and *Poa*-species) also occurs. As in all very windy habitats a variety of cushion-plants are characteristic of the subantarctic.

X The Alpine Vegetation of Mountainous Regions

1 Various Types of Upper Forest Limit or Timberline

Altitudinal belts commencing above the upper forest limit, irrespective of the absolute altitude, are termed alpine. In Middle and Central Asia, however, and in certain parts of the Andes, the mountains are so dry that a forest belt is entirely lacking. In such circumstances where, with increasing altitude, the alpine vegetation follows upon semidesert or steppe, the limits of the alpine belt are difficult to distinguish. In mountains of the Arctic zone the alpine belt and the tundra are one and the same.

The most thorough investigations have been carried out in the Alps, although the custom of using the alpine belt as summer pasture, practiced for over 1,000 years, has drastically changed the upper forest limit. A distinction can be made between:

1. *A theoretical, climatic, upper forest limit,* which is coincident with a definite contour line, but which the forest can never, in fact, quite reach because of such obstacles as walls of rocks, talus slopes, or cold air currents.

2. *A potential upper forest limit,* to which the forest would extend under natural conditions in the absence of human interference.

3. *The actual forest limit,* to which the forest extends at present. As a results of tree cutting and grazing in the Alps, this is usually 100–200 m lower than the potential limit, and, furthermore, the forest is greatly thinned out so that it was formerly thought necessary to distinguish between a forest limit and a tree line, the latter given by a line joining the solitary trees.

In the American Rocky Mountains, apart from places where ore was previously mined, completely natural conditions still prevail. The summits are often rounded, such as Pike's Peak, and it can be seen quite clearly that spruce forest consisting of *Picea engelmanii,* very similar to the European spruce, extends uninterruptedly up to 3,750 m. It is followed by a stunted spruce "Krummholz" zone reaching up to 3,800 m, which probably represents the climatic timber line.

This type of timber line can be found wherever, under humid conditions, shade-enduring species constitute the upper forest limit. Nevertheless, in mountainous regions with much snow, as for example on Mount Rainier (Washington) and in Utah, the forest, consisting of *Picea engelmanii* and *Abies lasiocarpa*, thins out at the timber line to form isolated groups of trees which, without exception, stand upon slight, mostly rocky, elevations, where the snow is not so deep and whence it probably melts earlier.

Similarly, the upper forest limit is much more interrupted where, under arid conditions, light-demanding species (which do not produce Krummholz) extend up to the alpine belt. This is the case with *Pinus hartwegii* in Mexico and probably holds also for the arid Mediterranean upper forest limit, although this has been disrupted everywhere by human interference.

Yet a third type is recognizable, where the upper forest limit consists of evergreen broadleaved species of subtropical origin, for example, on Mount Egmont in New Zealand, where *Nothofagus* is missing. Such species become shorter and shorter with increasing altitude and finally form a compact, closed canopy only 1 to 2 m above the ground. This is succeeded by a dense scrub of shrub-like Compositae *(Senecio, Olearia)* and *Hebe* (closely related to *Veronica),* dove-tailed in a mosaic-like fashion with the high-alpine tussock grassland. An upper forest limit is very difficult to distinguish.

A similar situation is found in the wet tropics, but more exact investigations have still to be carried out.

2 Climatic Conditions in the Alpine Belt

The mountains of the various climatic zones exhibit such a large variety of climatic differences that it is difficult to find a single common feature. This is perhaps only the decrease in mean annual temperature with increasing altitude. The intensity of radiation from the sun also increases steadily with altitude, but the total irradiation is not necessarily larger since, in certain mountainous districts (New Zealand, Paramo), the alpine belt is marked by heavy cloud and mist. Although as a rule precipitation increases with altitude, a rapid drop can occur in the alpine belt in the tropics if the cloud cover lies lower down, as is the case on Kilimanjaro. The Pico de Teide on

Teneriffa, too, rises high above the trade-wind clouds; at 2,367 m the rainfall amounts to only 369 mm, and, further up, the desert-like vegetation provides evidence of even drier conditions.

The more pronounced the seasonal temperature variations during the course of the year in the alpine belt, given adequate humidity, the more favorable for the plants, since they can exploit the relatively warm summer for growth and survive the winter in a state of dormancy. Conversely, in mountain regions of the equatorial zone the climate with merely diurnal variations in temperature is particularly adverse to plant growth (p. 57), and the forest limit is much lower than, for example, in the Himalayas. A continental climate with high summer temperatures resulting from the long periods of sunshine leads to a particularly high upper forest limit.

Such climatic differences render it impossible to deal with the vegetation of the alpine belt of all mountainous regions under one and the same heading. It cannot even be said of the plants that they are invariably low and that they always exploit the favorable temperature conditions of the microclimate on the ground. The sudden appearance of tree-like *Senecio* spp. and tall *Espletia* spp. in the high alpine belts of the tropics is difficult to explain ecologically.

As has already been pointed out in the introduction, it is more appropriate to deal with the vegetation of mountains under the vegetational zone to which they belong spatially, and, accordingly, the tropical mountains as well as those of the subtropics and the winter-rain regions have already been discussed, at least briefly.

At this point, only the alpine belt of the mountains of the temperate zone of the Northern Hemisphere will be dealt with, taking the Alps as a typical example.

3 Ecological Conditions Above the Timber Line in the Alps

The alpine belt and the Arctic tundra are often compared with each other. Although they both have a short period of vegetation, it must be emphasized that, apart from this, the alpine climate differs in almost every respect from that of the Arctic. Whereas the Arctic winter is perpetually dark and the summer continuously light, the length of day and night in the Alps is the same as in the adjacent lowland. The intensity of irradiation in the Arctic is weak, whereas in the Alps it increases steadily with increasing altitude and on clear winter days it is scarcely less than in summer. The daily temperature variations in the Arctic on clear summer days are very small but in the Alps very pronounced, on account of the intensive radiation during the day and the large amounts of outgoing radiation at night. At high altitudes the temperature differences between shady and sunny situation is extreme.

Contrasting situations also exist with respect to rainfall. In the Arctic the annual precipitation is very low, and the degree of wetness can be attributed to low evaporation and poor drainage. In the Alps, on the other hand, precipitation is very high but drainage is good, and in winter there is a thick covering of snow. Permafrost in the soil, a typical feature of the Arctic, does not play a role in the Alps, nor is solifluction so noticeable because rock and rubble habitats predominate at great altitudes and more, level ground with deep soil is rarely encountered.

Bearing these enormous climatic differences in mind, it is surprising to discover that many species occur both in the Arctic and in the Alps, as for example *Dryas octopetala, Saxifraga oppositifolia, Salix herbacea,* and so on. Nevertheless, these are probably different ecotypes with varying ecophysiological behavior. Transplantation experiments have not as yet been performed in order to test the truth of this supposition, although distinct arctic and alpine ecotypes have been shown to exist in *Oxyria digyna.*

As a result of the great depth of the snow blanket in the mountains, it is not the length of the vegetational season, as governed by the air temperature, which is of interest to the plants, but the period of time for which the ground is free of snow, known as the "aper" season (Latin, *apertus* = open). This depends to a large extent upon relief, wind direction, and exposure. Snow collects in the hollows and forms cornices on the lee side of ridges, but is blown off the windward side. If the latter is sunny as well, then the snow melts and the habitat is snowfree ("aper") all the year round. A shady, windward slope, however, is not warmed up by irradiation. In the presence of large drifts of snow at the foot of a slope facing north the snowfree season is reduced to a minimum on the so-called snow patches ("Schneetälchen") or is completely lacking, the snow remaining throughout the summer. The snowfree season can, however, vary in length from year to year on the same habitat according to the snowfall. The average length of the snowfree season decreases with increasing altitude and is theoretically zero at the climatic snowline. In individual cases, however, such as on steep rock faces, it can be very long, even above the snow line, and this is why flowering plants are found in the Alps in the nival belt, above the climatic snow line.

Climatic measurements have been made in the Ötztaler Alps at an altitude of 2,072 m above sea level over the entire course of five years. The climate is continental-alpine, and precipitation amounts to 830 mm, with rain on an average every second day. The longest rainless period lasted 14 days. The mean air temperature was below 0° C from November to March, with an absolute minimum of − 20° C, a maximum of 27.6° C, and an annual mean of 1.8° C. In contrast to the situation in the Arctic, the temperature on the ground beneath the covering of snow scarcely falls below zero. On the bare ground, however, extreme values of − 17.7° C and + 80° C were measured, although the latter value can only be attained by irradiation in wind-sheltered spots and on a dark soil of low heat conductivity. The effect of wind mitigates the extremes near the ground.

In any case, the microclimate, even on sunny days, is propitious with regard to temperature. If leaves are exposed to direct sunlight, their temperature can be as much as 22° C higher than that of the air. Every mountain climber is familiar with the warm niches which can be exploited by low, ground-hugging plants. In cloudy weather the temperature differences tend to be equalized.

From what has already been said it is clear that as far as the vegetation is concerned, there is no such thing as a standard climate in the alpine belt. Instead, it is split up into very small climatic units which can differ vastly from one another within a very short distance, as for example on the sunny and shady sides of a boulder. The way in which the snow is distributed in winter is of primary importance and must be known in order to be able to estimate the length of the snowfree season and so to understand the pattern of vegetation.

Roughly speaking, the vegetational belts above the upper forest limit, moving upward, take the following order:

The climatic upper forest limit

Lower alpine belt: (Succeeding each other in the order given) stunted trees, *Pinus montana* or *Alnus viridis* (Krummholz), dwarf shrubs.

Middle alpine belt: closed alpine meadow on level ground with an abundance of herbs.

Upper alpine belt: alpine mat patches which are in a state of gradual disintegration on account of solifluction.

The climatic snow line

Lower nival belt: alpine mat-patches, cushion-like plants common, patches of moss and lichen.

Middle nival belt: mosses and lichens predominate, interspersed with cushion-plants and rock-crevice plants.

Upper nival belt: Solitary flowering plants up to the upper limit, otherwise only moss and lichen.

Absolute altitudes cannot be given for the individual belts since all of the limits, including the upper forest limit and the climatic snow line, are 400 to 600 m higher in the central Alps than in the mountain chains to the north and to the south. This is due to the small amount of cloud and resultant larger total irradiation in the central Alps.

Temperature inversions and cold air pools play an important role in certain places. In the eastern Alps, near Lunz am See, a temperature of $-51°$ C has been recorded at an altitude of only 1,270 m in a doline from which the cold air cannot escape. Even in the height of summer outgoing radiation is accompanied by night frost, and trees are unable to grow on the floor of such dolines.

Another source of disruption of altitudinal belts is to be found in avalanches: In their wake the alpine vegetation continues far down into the forest belt. Even on the shallow, impoverished soil on dolomite, which undergoes little erosion, alpine exclaves can be found in the middle of the forest belt, and surviving habitats of alpine species in the bogs of the alpine foreland are familiar to the botanist. In habitats such as these the unpretentious but slowly growing alpine species are protected from the competition of other plants.

4 Ecophysiological Investigations on the Central European Alpine Species

The process of developing hardiness to frost takes a similar annual course in the dwarf evergreen shrubs to that already observed in the spruce (p. 202): Hardiness is developed in late autumn, and the process is reversed in spring. Despite their growing much higher up in the Alps than the spruce, the maximum degree of frost hardiness achieved by the dwarf shrubs is usually lower (less than $-30°$ C) since they are protected by a blanket of snow from the lowest winter temperatures. Only *Loiseleuria*, growing on windy habitats free of snow in winter, develops a larger degree of hardiness. Although the wind prevents the lowest temperatures from being reached, it enhances the danger of winter drought. If *Loiseleuria* is suspended in the open in winter, despite its xeromorphic structure, it dries out within 15 days. In its snowfree natural habitat, however, it normally grows tightly pressed to the ground, and the sun thaws any snow held between its shoots so that occasional water uptake is possible. The dwarf shrubs beneath their blanket of snow are not exposed to the dangers of winter drought.

The water budget is fairly well balanced in summer on account of frequent rainfall, and only when irradiation is intense or when a strong wind blows are the plants subjected to a large degree of evaporation for a few hours. The effects of wind are lessened near the ground. Even in habitats that appear dry on the surface, such as talus or scree slopes and rock faces, the soil carries abundant water. In this kind of habitat the plants develop an extensive root system or tap roots capable of penetrating into the rocky crevices, even though normally their root systems are very shallow and confined to the upper soil layers. The propitious water balance is reflected in the low cell-sap concentration of 8 to 12 atm, and even in xeromorphic species such as *Dryas*, *Carex firma*, and *Androsace helvetica*, it never rises to 20 atm. Here again, it would probably be more correct to speak of a peinomorphosis resulting from nitrogen deficiency than of xeromorphosis, since the uptake of nitrogen is, in fact, more difficult at low soil temperatures. Luxuriant, hygromorphic herbs are found only on *N*-rich habitats such as the resting-places of livestock.

If the total water lost in one year by the plant cover of alpine meadow communities is calculated, the figure of 200 mm is obtained. This is one-third of the summer rainfall and one-quarter of the total annual rainfall, even in the relatively dry central Alps. Evaporation depends, above all, upon the wind and is for this reason influenced by the relief, although in the inverse direction to snow deposition.

The short growing season in the alpine belt gives rise to the question of adequate production. The days are shorter than in the Arctic, but the light intensity is stronger and nocturnal temperatures and losses due to respiration are lower. Under favorable conditions of illumination 100 to 300 mg CO_2/dm^2 can be assimilated in one day. One month of good weather suffices for the accumulation of sufficient reserves for the coming year and for the ripening of seeds, but since the growing season last three months adequate production is in any case ensured (20).

Primary production in plant communities depends largely upon the density of vegetation. The following values were found:

Closed meadows	50–276 g/m²
Dryadeto-Firmetum	91 g/m²
Salicetum herbaceae	85 g/m²
Oxyrietum	15 g/m²
On limestone scree	1 g/m²

Photosynthesis is less intense in dwarf shrubs than in herbaceous species, but since the total leaf area of the former is larger and the period of vegetation longer in the lower alpine belt, they achieve a higher primary production. The most unfavorable conditions of all are met with on the snow patches ("Schneetälchen") on the north-facing slopes in the siliceous rock regions. The snow melts very slowly from the edges of such patches, and the ground gradually becomes exposed. This means that within a very small area a series of zones can be recognized, in which the ground is free of snow ("aper") for ever-shorter periods. The soil of such habitats is rich in humus, slightly acid, and invariably well-provided with water from the melting snow, but for this reason also relatively cool. If the snowfree period lasts for three months, mats of *Carex curvula* develop. If the growth season is reduced to 2 months, the willow *Salix herbacea* predominates, a woody species of which only the tips of the shoots are above ground so that its leaves form a compact mat. *Salix herb-*

acea only bears fruit after a winter with particularly little snow when the snowfree season amounts to 3 months. There is also a scattering of very small plants such as *Gnaphalium supinum, Alchemilla pentaphylla, Arenaria biflora, Soldanella pusilla, Sibaldia procumbens*, etc. Given an even shorter period for growth, only mosses manage to grow because they require no time for the production of flowers and fruits. The most common of these is *Polytrichum sexangulare (P. norvegicum)*, and if the snowfree season is too short for such green mosses, then only a liverwort, *Anthelia juratzkana*, is found. It looks like a mouldering crust. This latter zone does not become snowfree at all if there has been an unusually large amount of snow in the preceding winter.

The vegetational zones on the snow patches ("Schneetälchen") give us an idea of what the altitudinal belts in the mountains would look like if there were large expenses of level ground and if the snowfree season steadily decreased with increasing altitude up to the climatic snow line. Although the upper slopes in the Alps are too steep, there is flat ground in the Scandinavian mountains where Salicetum herbaceae is found, and, in places where the snowfree period lasts for an average of one month, a Polytrichetum with *Anthelia juratzkana*. Near the snow line, however, development of the vegetation is disrupted by solifluction and the formation of polygonal soil structures. On the permanent snow the last sign of life is provided by the alga *Chlamydomonas nivalis*, which lends a rosy hue to the surface of the snow.

Since bare rock predominates in the alpine belt of the Alps, its chemical composition is of great significance for the vegetation because it governs the reaction of the soil. *The floristic differences between the calcareous Alps and the crystalline siliceous central Alps are very pronounced.* Accordingly, a distinction is made between limestone-demanding or *basophile species* and *acidophile species* that avoid limestone. Vicarious species like the alpine rose are often encountered: *Rhododendron hirsutum* on limestone, *Rh. ferrugineum* on siliceous rock or acid humus soil.

The conditions which have been discussed here hold not only for the Alps but for all the mountains of the temperate climatic zone of the Holarctic floristic realm, if characterized by a similar level of precipitation. There are only certain floristic differences. In the mounains of the Southern Hemisphere, on the other hand, often with very different climatic conditions, the floristic differences are so great that despite certain common features the vegetational pattern cannot be compared with that of the Alps.

Summary

Having discussed the individual vegetation zones, we shall now turn to a summary of the phytomass and yearly primary production in the whole biosphere. The *biosphere* is that thin layer at the earth's surface in which living organisms exist and where, therefore, biological cycling takes place. It includes the upper horizons of the soil in which plants root as well as the atmosphere near the ground, insofar as organisms penetrate this space, and all the surface waters.

More than 99 percent of the earth's biomass is phytomass, to which we shall limit our discussion. Amounts of phytomass are distinctly related to vegetation zones.

Because accurate determination of phytomass and primary production is difficult, we have only had gross estimates available. However, in 1970 Bazilevich, Rodin, and Rozov (22) published in Russian more accurate calculations, based on the rapidly accumulating literature, for the various thermal zones and bioclimatic regions of the earth. Their point data were extended to an areal basis using the maps of vegetation and soils prepared for the Soviet Physical-Geographical Atlas of the World (1964).

They used five thermal zones: (1) Arctic, (2) Boreal, (3) Temperate, (4) Subtropical, (5) Tropical. The first two zones have humid climates; the three others were divided into (a) humid, (b) semiarid, and (c) arid (cf. the map in Fig. 79).

This classification differs from the vegetation zones used in this book as follows:

Thermal zones and climatic regions of Bazilevich, Rodin, and Rozov:		Our vegetation zones:
Zone 1	corresponds to	Zone IX
Zone 2	corresponds to	Zone VIII
Zone 3 (a–c)	corresponds to	Zone VI and VII

Zone 4 (a–c) corresponds to Zone V, IV and III
(poleward of $23^1/_2°$ latitude,
the Tropics of Cancer and
Capricorn)

Zone 5 corresponds to Zone I, II, and III
(equatorward of $23^1/_2°$
latitude)

Dry weights are given in metric tons (1,000 kg), the average phytomass and the average yearly primary production in t/ha. The areas of the different regions do not include areas of rivers, lakes, glaciers, and permanent snow. Areal summation gives the phytomass and production of the land surface of the earth. In the following table organic production of the water areas is also included. The figures represent potential values, that is, for natural, not man-modified, vegetation.

Distribution of potential production on the earth.* After Bazilevich, Rodin, and Rozov 1970 (22).

Climatic zones		Area (in 10^6 km²)	Phytomass		Primary production	
			Total (in 10^9 t)	Average (in t/ha)	Total (in 10^9 t/yr)	Average (in t/ha·yr)
Polar		8.05	13.8	17.1	1.33	1.6
Boreal		23.2	439.	189.	15.2	6.5
Temperate	h	7.39	254.	342.	9.34	12.6
	s	8.10	16.8	20.8	6.64	8.2
	a	7.04	8.24	11.7	1.99	2.8
Sub-tropical	h	6.24	228.	366.	15.9	25.5
	s	8.29	81.9	98.7	11.5	13.8
	a	9.73	13.6	13.9	7.14	7.3
Tropical	h	26.5	1166.	440.	77.3	29.2
	s	16.0	172.	107.	22.6	14.1
	a	12.8	9.01	7.0	2.62	2.0
Land area		133.0	2400.	180.	172.	12.8
Glaciers		13.9	0	0	0	0
Lakes & rivers		2.0	0.04	0.2	1.0	5.0
Oceans		361.0	0.17	0.005	60.0	1.7

* The original gives more significant figures. They are rounded to 3 here to avoid the impression of too great accuracy.

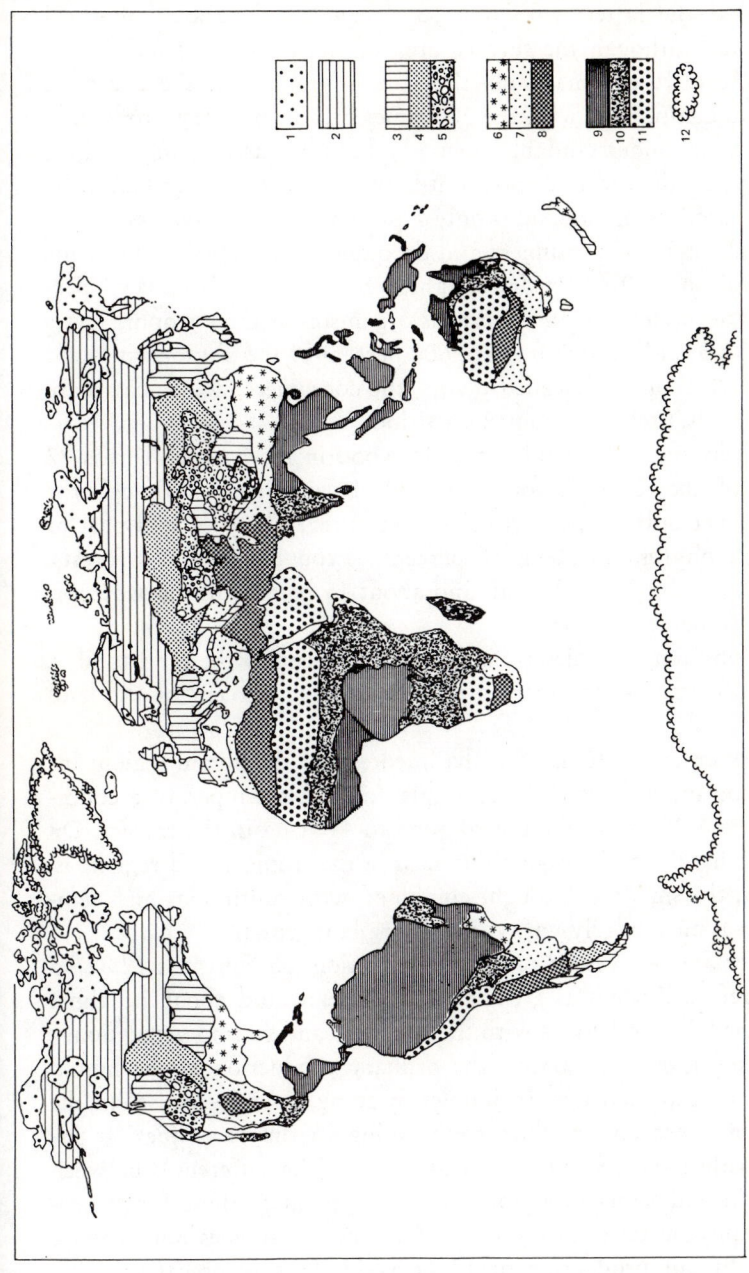

Fig. 79. Thermal zones and bioclimatic regions. From Bazilevich, Rodin and Rozov 1970. 1. Arctic zone, 2. Boreal zone, 3—5. Temperature zone, 3. humid region, 4. semiarid region, 5. arid region, 6—8. Subtropical zone, 6. humid region, 7. semiarid region, 8. arid region, 9—11. Tropical zone, 9. humid region, 10. semiarid region, 11. arid region, 12. Glaciers and permanent snow.

A comparison of terrestrial production with that in the oceans shows that the latter, with 60×10^9 t, equals only about one-third of the land although the surface area of the oceans is almost three times larger. It is remarkable that the phytomass in the oceans is minute in comparison with the 300 times greater primary production there. This is understandable since plankton plants are single-celled, and they divide and multiply continuously. On the other hand, the primary production on land is only 7 percent of its phytomass.

For the mass of consumers and decomposers on all the continents a figure of only 20×10^9 t dry weight is given. This is less than 1 percent of the phytomass. In the oceans organisms of these trophic levels amount to 3×10^9 t, or more than 600 percent of the phytomass found there. Instead of single-celled plants, the consumers in the oceans are large animals, useful for human nutrition.

The phytomass on land is mostly wood in the forests. This is 82 percent of the total phytomass on all the continents although the forests cover only 39 percent of the land area. The principal part of the forest phytomass, about 50 percent, is found in tropical forests, about 20 percent in the boreal, and about 15 percent in the subtropical and temperate 3 zones.

The phytomass in deserts is only 0.8 percent of the land total, a very small amount considering they occupy 22 percent of the land area.

The average phytomass in t/ha in forests of the humid regions increases continuously with increasingly favorable temperature conditions from 189 t/ha in the boreal zone to 440 t/ha in the tropics. On the other hand, the average phytomass in the tropical arid regions is, at 7 t/ha, the smallest. Drought combined with continuous high temperatures is particularly unfavorable for plant growth.

As for average yearly primary production, on land it is 12.8 t/ha and therefore 7 times as great as in the oceans and about 2.5 times that in the lakes and rivers with their aquatic and swamp vegetation.

As one nears the equator, the primary production of the humid regions on land increases. It doubles in going from the boreal to the temperate zones and doubles again going to the subtropics. It increases little from the subtropics to tropics. The differences between the humid and semiarid regions are not so great as those for the respective phytomasses since the wood in the forest does not produce dry weight but production of 13.8 t/ha in the subtropical semiarid regions is noteworthy. It depends on the often very luxuriant and productive ephemeral vegetation which can develop during the favorable, cooler rainy part of the year.

The total, yearly potential primary production of the biosphere on the land, in the oceans, and in lakes and rivers is, according to today's best knowledge, 233×10^9 t/ha·yr.

Notes

1. WALTER, H./STRAKA, H.: Arealkunde. Floristisch-historische Geobotanik. Einführung in die Phytologie, Vol. III, Part 2, Stuttgart 1970, 2nd Edition.

2. See also WALTER, H. and KREEB, K.: Die Hydration und Hydratur des Protoplasmas der Pflanzen und ihre öko-physiologische Bedeutung. Protoplasmatologica II/C/6, Vienna 1970.

3. In the equatorial zone the temperature is the same throughout the entire year. The curve for potential evaporation runs parallel to that for air humidity and is a mirror image of the rainfall curve.

4. WENT, F. W., and STARK, N.: Mycorrhiza. Bio Science 18, 1035—1039, 1968.

5. ASPREY, G. F., and ROBBINS, R. G.: The vegetation of Jamaica. Ecol. Monogr. 23, 359—412, 1953.

6. In Malaya the old leaves drop after the new ones have appeared if the weather is wet, but beforehand if the weather is dry. In this manner deciduous woody species have arisen in a climatic zone with a drought season. See HOLTTUM, R. E.: Evolutionary trends in an equatorial climate. Symp. Exper. Biol. No. VII. Evolution, 1953.

7. ZELLER, O.: Blütenentwicklung und Ausläuferbildung bei Fragaria ananasa Duch. in verschiedenen Höhenlagen der Insel Ceylon. Angew. Bot. 43, 159—173, 1969.

8. COUTINHO, L. M.: Untersuchungen über die Lage des Lichtkompensationspunktes einiger Pflanzen zu verschiedenen Tageszeiten mit besonderer Berücksichtigung des „de-Saussure-Effektes" bei Sukkulenten. Beiträge zur Phytologie (WALTER-Festschrift) in Arbeiten d. Landw. Hochsch. Hohenheim, Vol. 30, Stuttgart 1964.
COUTINHO, L. M.: Novas observações sôbre a ocorrência do „Efeito de de Saussure" e suas relações com a suculência, a temperatura folhear e os movimentos estomaticos. Bol. 331, Fac. Fil., Ciênc. e Letr. da Univ. São Paulo, Botan. 24, 77—102, 1969.

9. WALTER, H., and MEDINA, E.: Die Bodentemperatur als ausschlaggebender Faktor für die Gliederung der subalpinen und alpinen Stufe in den Anden Venezuelas. Ber. Dtsch. Bot. Ges. 82, 275—281, 1969.

10. MÜLLER, D., and NIELSEN, J.: Production brute, pertes par respiration et production nette dans la forêt ombrophile tropicale. Forstl. Forsgsv. in Denmark 29, 69—160, 1965.

11. MEDINA, E.: Bodenatmung und Streuproduktion verschiedener tropischer Pflanzengemeinschaften. Ber. Dtsch. Bot. Ges. 81, 159—168, 1968. The values given by E. D. SCHULZE (Ecol. 48, 652—653, 1967) for Costa Rica are, according to their author, 10 times too high; they are to be reduced by a factor of 10 and then agree with those given by MEDINA. WANNER (J. Ecol. 58, 543—547, 1970) was also brought up short by SCHULZE's high results, and he questioned them.

12. On the east coast of India the rain-bringing monsoon blows during the cool season so that the summers are dry; woody species with evergreen, xeromorphic leaves predominate.

13. In Africa too, palm-savannas with *Hyphaene* and *Borassus* are widespread on alternately wet and dry soils with a grassy cover, which is often burned.

14. SCHOLANDER, P. F., HAMMEL, H. T., BRADSTREET, E. D., and HEMMINGSEN, E. A.: Sap pressure in vascular plants. Science 148, 339—346, 1965. SCHOLANDER, P. F.: How mangroves desalinate seawater. Physiol. Plant. 21, 251—261, 1968.

15. WOODEL, S. R. J., MOONEY, H. A., and HILL, A. J.: The behavior of *Larrea divaricata* (creosote bush) in response to rainfall in California. J. Ecol. 57, 37—44, 1969.

16. LANGE, O. L.: Die funktionellen Anpassungen der Flechten an die ökologischen Bedingungen arider Gebiete. Ber. Dtsch. Bot. Ges. 82, 3—22, 1969. LANGE, O. L., KOCH, W., and SCHULZE, E. D.: CO_2-Gaswechsel und Wasserhaushalt von Pflanzen in der Negev-Wüste am Ende der Trockenzeit. Ber. Dtsch. Bot. Ges. 82, 39—61, 1969.

17. FLOWERS, S., and EVANS, F. R.: The flora and fauna of the Great Salt Lake region, Utah. In Salinity and aridity; Monographiae Biologicae 16, 367—396, The Hague 1966.

18. DOLEY, D.: Water relations of *Eucalyptus marginata* SM under natural conditions. J. Ecol. 55, 597—614, 1967. See also HELLMUTH, E. O.: Eco-physiological studies on plants in arid and semi-arid regions in Western Australia. I. Autecology of *Rhagodia baccata* (Labill.) Mog. J. Ecol. 56, 319—344, 1968; II. Field physiology of *Acacia craspedocarpa* F. Muell. J. Ecol. 57, 613—634, 1969; and III. Comparative studies on photosynthesis, respiration and water relations of 10 arid zone and 2 semi-arid zone plants under winter and late summer climatic conditions. J. Ecol. 59, 225—259, 1971.

19. HOFMANN, W.: Geobotanische Untersuchungen in Südost-Spitzbergen 1960. In Erg. d. Stauferland-Exp. 1959/60, Issue 8, Wiesbaden 1968.

20. MOSER, W.: Die Photosyntheseleistung von Nivalpflanzen. Ber. Dtsch. Bot. Ges. 82, 63—64, 1969.

21. In P. VILA (Geografia de Venezuela II, Caracas 1965), on page 140, the statement is to be found that Dr. ROBERT TSCHUDY, on the basis of many pollen analyses, was unable to detect any appreciable change in the vegetation of the llanos on the Orinoco since the Tertiary Age, i. e. they were not forested areas in earlier times.

22. BAZILEVICH, N. I., RODIN, L. E., ROZOV, N. N.: Geograficheskiye Aspekty izucheniya biologicheskoy produktnivosti. Materialy V syezda Geograficheskovo obschestva Soyuza SSR. Leningrad 1970. Translated as „Geographical aspects of biological productivity" in Soviet Geography, Rev. & Transl. 12 (5): 293—317, 1971. On the Soviet Physical-Geographical Atlas of the World, see Soviet Geogr., Rev. & Transl. 6 (5/6): 1—403, 1965 and Ecology 48 (2): 328—9, 1967.

Sources of Illustrations

The following illustrations have been taken from other works of the author, and are in most cases reduced in size:

Figs. 3, 6, 7, 18, 30, 35, 39, 45, 64, 65, 70 from H. WALTER: Standortslehre (Phytologie Vol. III/1). Verlag Eugen Ulmer, Stuttgart 1960, 2nd Edition.

Figs. 4, 5, 12, 13, 29, 51, 53, 61, 63, 66, 71, 72, 74, 76 from H. WALTER: Die Vegetation der Erde in öko-physiologischer Betrachtung, Vol. I, 2nd Edition 1964; Vol. II, 1968. VEB Gustav Fischer Verlag, Jena.

Figs. 36, 37, 38, 42 from H. WALTER and O. H. VOLK: Grundlagen der Weidewirtschaft in Südwestafrika. Verlag Eugen Ulmer, Stuttgart 1954. All climate diagrams in Figs. 8, 9, 17, 25, 41, 49, 52, 55, 56, 59, 60, 62, 73, 75, 78 from H. WALTER and H. LIETH: Klimadiagramm-Weltatlas. VEB Gustav Fischer Verlag, Jena 1967.

Photographs: Figs. 16, 24, 27, 28, 31, 33, 44, 48, 54, 57, 67, 68, 69, 77 by E. WALTER; Figs. 46, 47, 50, 58 by H. WALTER; Fig. 20 by J. SCHMIDT; Fig. 21 by H. SCHENK.

Subject Index